Reading *Ḥayy Ibn-Yaqẓān*

Reading *Ḥayy Ibn-Yaqẓān*

A Cross-Cultural History of Autodidacticism

AVNER BEN-ZAKEN

The Johns Hopkins University Press
Baltimore

© 2011 The Johns Hopkins University Press
All rights reserved. Published 2011
Printed in the United States of America on acid-free paper
2 4 6 8 9 7 5 3 1

The Johns Hopkins University Press
2715 North Charles Street
Baltimore, Maryland 21218-4363
www.press.jhu.edu

Library of Congress-Cataloging-in-Publication Data

Ben-Zaken, Avner.
Reading Ḥayy Ibn-Yaqẓān : a cross-cultural history of autodidacticism / Avner Ben-Zaken.
p. cm.
Includes bibliographical references and index.
ISBN-13: 978-0-8018-9739-9 (hardcover : alk. paper)
ISBN-10: 0-8018-9739-4 (hardcover : alk. paper)
1. Ibn Tufayl, Muhammad ibn 'Abd al-Malik, d. 1185. Risalat Ḥayy ibn Yaqẓān.
2. Islamic philosophy. I. Title.
B753.I54B46 2010
181'.92—dc22   2010008071

A catalog record for this book is available from the British Library.

Special discounts are available for bulk purchases of this book. For more information, please contact Special Sales at 410-516-6936 or specialsales@press.jhu.edu.

The Johns Hopkins University Press uses environmentally friendly book materials, including recycled text paper that is composed of at least 30 percent post-consumer waste, whenever possible. All of our book papers are acid-free, and our jackets and covers are printed on paper with recycled content.

To
Atar, Michaella, and Eleanor

At first who invented any art whatever that went beyond
the common perceptions of man was naturally admired by men,
not only because there was something useful in the inventions,
but because he was thought wise and superior to the rest.
—Aristotle, *Metaphysics*

What the mind thinks must be in it just as characters may be said
to be on a non-written table.
—Aristotle, *De Anima*

I am not bound to swear allegiance
to the word of any master
I am carried, a guest, wherever the storm-wind blows me.
Trying to bend world to self, and not self to world.
—Horace, *Epistles*

To teach a little child is like writing on a blank slate.
—Alisha Ben-Avuya, "Avot," *Mishna*

His self-knowledge was Himself.
—Abū Baker Ibn-Tufayl, *Ḥayy Ibn-Yaqẓān*

Man has begun to be his own master, let him use his judgment!
—Francis Bacon, *The Great Instauration*

Let us then suppose the mind to be, as we say, white paper, void of all
characters, without any ideas. How comes it to be furnished? . . .
From Experience.
—John Locke, *Essay Concerning Human Understanding*

CONTENTS

Preface xi

INTRODUCTION. The Pursuit of the Natural Self   1

1 Taming the Mystic: *Marrakesh, 1160s*   15

2 Climbing the Ladder of Philosophy: *Barcelona, 1348*   42

3 Defying Authority, Denying Predestination, and Conquering Nature: *Florence, 1493*   65

4 Employing the Self and Experimenting with Nature: *Oxford, 1671*   101

CONCLUSION. Sampling the History of Autodidacticism   126

Notes   139
Essay on Sources   171
Index   183

PREFACE

M Y INTEREST IN THE CIRCULATION and construction of autodidacticism in the early modern period grew out of reconsiderations of fundamental concepts of modernity that arguably marked a sharp break between European culture and that of the rest of the world. My previous book *Cross-Cultural Scientific Exchanges in the Eastern Mediterranean, 1560–1660* (JHUP, 2010) focuses on the rise of heliocentrism and the revolutionary displacement of man from the center of the cosmos. Heliocentrism marked the first European scientific achievement that broke Europeans away from the scientific traditions of the Near East, arguably setting them on a linear path of development and shifting them from the medieval preoccupation with relations of man with God to the new anxiety concerning the relations of the individual and the cosmos. In that volume I show, however, that after the Copernican Revolution cross-cultural practitioners considered Near Eastern sources as encapsulating hints about heliocentrism, and they further engaged with the question of the relations of man and the cosmos by invoking, reconstructing, and corroborating myths of ancient theology.

In the current book, I explore and reassess yet another fundamental concept of modernity—autodidacticism—which not only represented the rejection of traditional intellectual authorities but also marked the rise of European experimentalist practices that diverged from transcendent medieval authorities. In tracing the circulation and reception of a quintessential autodidactic text, *Ḥayy Ibn-Yaqẓān*, within various late medieval and early modern European intellectual cultures, the book at hand demonstrates how autodidacticism and experimentalism took shape along cross-cultural exchanges in which this quintessential autodidactic medieval text was translated, modified, and adopted to particular historical moments.

In a way, this project represents a world history of a single book, a cross-cultural account of the reading of one specific text. It plots the many readings of this text through different cultural and historical contexts, each driven by cultural questions specific to that context. As the following chapters suggest, local readings of the story left traces in society and culture, transforming forms of sociability, allowing any individual to fashion his own life, challenging social orders previously understood as unchangeable, and permitting new modes of thought that relied not any longer on transcendent authorities but rather on firsthand experience. Such traces ramified to politics, changing people's relationships with power, collecting and transforming what had been scattered individual goodwill and firsthand experience into participatory forms of politics.

Tackling this project was an enormous methodological and intellectual challenge. It required the exploration of four different time periods situated in four different places usually kept apart by artificial professional configurations and field divisions in the discipline of history. I suspect, then, that the book may appeal to scholars of history who work in divergent fields, for whom one chapter or another may be especially pertinent to their work. However, the book also represents a historiographic proposal for a more unified cultural history, interdisciplinarily fusing seemingly mutually exclusive fields. I do, therefore, hope that the book as a whole will be of interest to many scholars, no matter how far apart their interests may be.

This book was written during my term as a junior fellow at the Harvard Society of Fellows. There I was fortunate to experience intense intellectual exchange between scholars from various fields, breaking down disciplinary fences and encouraging me to pursue a multidisciplinary and cross-cultural approach to the history of autodidacticism.

Representing various vantage points, Bernard Bailyn, Eric Nelson, Nur Yalman and Amartya Sen frequently challenged my arguments over Monday dinners. Elizabeth Camp, Debra Cohen, David Elmer, Michael Gordin, Scott Johnson, Vanessa Ryan, and Olivier Tinland expanded my perspective through an extended period of friendly exchanges.

Because each chapter deals with far-removed themes, places, and time periods, I took it as a habit to exchange information, sources, and arguments with the leading scholars of each particular field. Lenn Goodman and Dimitri Gutas supplied useful comments about Ibn-Tufayl and twelfth-century Marrakesh. Herbert Davidson, Gad Freudenthal, and Moshe Halbertal assisted in embedding the story of Narbonni in Barcelona within the wider context of late medieval Jewish philosophical controversies. Darrell Rutkin, Brian Copenhaver, Arthur Lesley, and Mauro Zonta provided fruitful directions for sorting out the identity of Pico della Mirandola's translator as well as his writing strategy during his time in Florence. Mario Biagioli, Katherine Park, and Steven Shapin gave valuable observations regarding the connection between utopian societies and experimentalism. Jennifer Simchowitz introduced me to contemporary discussions in the Art of Sampling and stimulated me to apply it to historiography, proposing historical sampling as the historiographical framework for the book.

Ann Blair, Natalie Zemon-Davis, Carlo Ginzburg, Antony Grafton, Simon Schaffer, Theodore Porter, and Norton Wise oversaw the book project, followed its progress, and supplied invaluable encouragement.

I am indebted to the Milton Fund of Harvard University, which supported my work. Diana Morse at the Society of Fellows enormously supported my efforts and facilitated the project for some years.

I am also grateful for the professional assistance of librarians and archivists from various institutions around the world: the Houghton Library at Harvard University, the Bodleian Library at Oxford University, the University Library of Genoa, the Bavarian State Library in Munich, and the Biblioteca Comunale Teresiana in Mantua.

Hillel Eyal and Eugene Sheppard read the text and drew my attention to unclear passages, stimulating me to look for new strategies to present my arguments. Michal Lemberger edited the last draft and significantly improved the flow and clarity of the text.

Last, I would like to thank my family. My wife, Atar, and our daughters, Michaella and Eleanor, grounded my scholarly efforts in earthly concerns and provided the healthy perspective I needed to wrap up the project and prepare it for publication. This book is dedicated to them.

Reading *Ḥayy Ibn-Yaqẓān*

INTRODUCTION

The Pursuit of the Natural Self

WHAT WOULD YOU DO IF you were stranded on a desert island? This dilemma and its derivatives somewhat epitomize the almost obsessive modern pursuit for the "true inner self." In our shared cultural imagination, the desert island and the life of solitude allow one to actively explore himself and his surroundings, to experience life firsthand rather than mediated by culture, and to perfectly situate himself in his proper natural place. Self-discovery becomes essential to the purpose of modern life—to the pursuit of happiness. Without social and historical constraints, without the burden of tradition, and without false needs imprinted on our minds by cultural hegemony, the individual can trust firsthand experience to bring natural, and so complete, happiness.

Modern culture has celebrated the discovery of the "natural self" through stories of wilderness and solitude. Early in the eighteenth century, Daniel Defoe published *Robinson Crusoe,* one of the first English novels to establish a genre that strips human beings of their "cultural clothing" and situates them in nature, where they can access a universal self. Cast on a desert island, Robinson Crusoe's struggle for survival actually represented a process of self-discovery, of human resourcefulness, of active exploration leading to the creation of an "industrious virtuoso." Crusoe takes charge of his destiny and works to supply his needs, and through the application of autodidactic philosophical principles he observes that "as Reason is the Substance and Original of the Mathematicks, so by stating and squaring everything by Reason, and by making the most rational Judgment of things, every Man may be in time Master of every mechanick Art."

The genre of self-discovery in nature and solitude also took an educational turn. In the late nineteenth and early twentieth centuries, stories of wild boys were celebrated in popular culture. They represented a radical version of self-education through the image of the perfectly clean *tabula rasa* (blank slate): the wild boy living in nature, socially interacting only with animals and relying completely on his own natural instincts. Rudyard Kipling's *The Jungle Book* tells a story of a wild boy, Mowgli, lost by his parents in the Indian jungle during a tiger attack and adopted by wolves. Despite his upbringing, Mowgli possesses the power of dominion over beasts. His almost superhuman talents are ascribed to his upbringing—raised in the jungle by wild animals.

Mowgli represents a source of inspiration for another popular early twentieth-century hero. Edgar Rice Burroughs's *Tarzan* recounts the remarkable exploits of young Lord Greystoke, who is raised by a family of apes in the African jungle and becomes the worthy opponent of the jungle's most feared predators. The tale describes Tarzan and his relationship with the great apes, their natural enemies, and Tarzan's peculiar abilities as an autodidact. Natural instincts of shame and blushing differentiated him from animals. After observing that only animals went naked he covered himself, and "no longer did he feel shame for his hairless body or his human features, for now his reason told him that he was of a different race." Like the biblical Adam, Tarzan's first conscious observation was his own nakedness.

Tarzan and Mowgli soon became heroes in a series of children's books and movies, holding up isolated wild boys as modern cultural icons. In testing human nature in a natural laboratory where the initial conditions are created not by civilization but by nature, they symbolized the self-discovery of knowledge. Autodidacticism, however, is not the creation of the modern world, nor is it a modern novelty.

Ḥayy Ibn-Yaqẓān (Alive Son of the Vigilant) is the tale of the quintessential autodidact. A medieval philosophical treatise in literary form, written by the Andalusian philosopher Abū Bakr Ibn-Tufayl in the 1160s, it relates the story of human knowledge as it rises from a blank slate, through practical exploration of nature, to a mystical or direct experience of God. Its central argument is that human reason can independently access scientific knowledge unaided by religion or society and its conventions, leading not only to the tenets of natural philosophy but also to the attainment of mystical insight, the highest form of human knowledge.

The story takes place on Wāqwāq, an equatorial island in the Indian Ocean uninhabited by human beings, where Ḥayy had either been spontaneously generated when the elements on the island reached equilibrium or, more prosaically, had been born elsewhere of a forbidden marriage and taken by a watery trip in a box to end up on this no-place island. A gazelle hears Ḥayy's cries, suckles him, protects him, and takes care of him. At this early age he learns to imitate other animals' speech and has covered parts of his body with leaves after noticing the same parts covered with hair or feathers on the animals around him. He starts making tools to help him fight animals whom he considers dangerous and to more efficiently collect and supply food for himself. Having mastered self-preservation and self-subsistence, he uses the tools he has invented to protect other animals from predators or to save them in dangerous circumstances. Ḥayy is thus able to rule the island and become its superior creature.

The gazelle's death, when Ḥayy was only seven years old, directed his energies to the exploration of nature and himself. He begins by dissecting the body of his mother-gazelle to understand the reasons for her death. Following the trail of circulating blood, he finds his way to the heart and discovers in it two rooms, one with congealed blood and the other empty, from which he deduces that the body held a soul that departed it. In dissecting dead bodies of wild beasts he accurately inquires into living bodies, stretching his thought beyond the immediate utilitarian knowledge, until he becomes the greatest natural philosopher.

After exploring the life sciences, Ḥayy moves on to physics. Through trial and error he perceives that warm bodies grow cold and cold ones become hot; he watches water turn to steam, and steam to water; burning things to embers, ashes, flame, and smoke. When rising smoke is trapped in the hollow, he observes, it precipitates and in its place appear bits of solid matter, rather like earth. After discovering fire he starts experimenting with burning bodies. Seeing that fire always leans upward and is actively ascending, he persuades himself that it is a heavenly substance, and he tries its force on all things, throwing them into it, on which fire prevails more or less, according to the disposition of that body, as it is more or less fit to be kindled. Following this line of thinking, he understands that through practical exploration of the faculties of fire and light, the motion of air and the extension of water, man not only can discover the fundamental laws of nature but can also set himself apart from all animal species. He sees that he was created as a superior creature with a different end from all the rest, dedicated to a great task of knowing, naming, and conquering nature.

Ḥayy next approaches astronomy, observing the planets and stars, gazing on the orbits of the moon and the sun, finally deducing the contours of cosmology and correctly predicting the revolutions of the planets, classifying astronomy as a science that can be described and demonstrated by mathematics. He moves on to metaphysics, searching for the prime mover that put the sphere of the stars into motion, ascribing it to the "cause of itself" or, as he calls it, the "necessary being," the initiation of the eternal motion of the heavens. Not satisfied with this fundamental discovery, he searches beyond knowledge of the existence of a necessary being to experience direct communion with it.

Ḥayy then passes through three stages of embodied and active practice: he first imitates the plants, and then, by circling the island in harmony with the direction of the circulation of the heavens, the planets; finally, he confines himself in a cave to meditate, looking deeply into himself to gain communion with the necessary being, with God. There, he not only discovers the ultimate secrets of the cosmos but gains an ecstatic experience and complete felicity.

Later, two men, Salman and Absal, sail to Ḥayy's island. They find him and quickly realize they are dealing with an extraordinary, self-taught philosopher. Taking him back to their own island, they introduce him to other philosophers. Ḥayy quickly learns their language and engages them in discussions on philosophy and theology. After this encounter with society he understands the human condition, seeing that most men are no better than unreasoning animals, submissively accepting all the most problematical elements of the tradition and avoiding originality and innovation.

Though Ḥayy perceives himself to be intellectually superior, he feels unchallenged and bored, and so he goes back to his island, choosing the life of solitude over civilization, this time with Absal in tow. By the end of the story, Ḥayy and Absal have established a quasi-scientific society for the exploration of nature, the practice of meditation, and the achievement of perpetual felicity. In sum, *Ḥayy Ibn-Yaqẓān* tells the story of an isolated wild boy, an autodidactic prodigy, who, without parents, teachers, or language, takes control over nature through the use of instruments, discovers the laws of nature through practical exploration and experimentation, and achieves complete felicity through mystical mediation and communion with God.

Such pedagogical principles had already popped up in the ancient Sanskrit tales of Vishnu Sarma—*Panchatantra* (translated into manifold languages; in English, *Fables of Bidpai*), a collection of animal fables that functioned as indoctrination devices to condition the moral behavior of children. The fables stressed and valued interactive, voluntary, dynamic, reflective, open, frustrating, and risky learning over simplified, fixed, and authoritative knowledge.

Thus the story of Ḥayy was not the first of such pedagogical treatises, nor would it be the last. Despite the large historical chasm between the animals of the ancient *Panchatantra,* the medieval Ḥayy, and modern heroes such as Tarzan and Mowgli, they all place at center the sensations children experience in nature. Philosophical trends that arose in the early Enlightenment invoked such latent metaphors of children, stressing the *tabula rasa*—the blank slate on which experience and perceptions are impressed to create knowledge—and the value of self-teaching, to promote the new methodologies of empiricism and experimentalism as the best ways to advance human knowledge.

During the late seventeenth century, John Locke proposed childhood prodigies and autodidacts as the new icons of reason. In his *Essay Concerning Human Understanding* (1690) and *Some Thoughts Concerning Education* (1692), he introduced a mechanical epistemology of sensations and reflections based on the understanding of the mind as *tabula rasa*. Locke was the first to apply this epistemology to education, giving allowance to the rise of the image of the wondering boy, the perfect *tabula rasa*, as an alternative to the old wise man as the icon of wisdom.

Locke's essays provide the fundamental principles for the study of human development and the crucial formative effects of education. "Nine men out of ten," he holds in his essay *Some Thoughts Concerning Education*, reveal themselves, "good or evil, useful or not, by their education." For Locke, even all-but-insensible impressions on tender minds have important and lasting consequences, since a child's mind is a blank slate, lacking innate ideas. Knowledge can be achieved only through sensation and reflection. Children pick up their ideas of colors and tastes through observation and experience and then develop a new set of ideas by reflecting upon those sensations and arriving at the intellectual states of doubt, belief, reason, knowledge, will, and all other acts of mind. Locke's *Thoughts Concerning Education* portrays the ideal type as "a mind free, master of itself and all its actions" who, through constant practice, fashions his own world.[1]

As he notes in his earlier *Essay Concerning Human Understanding*,

> Our knowledge of substances is to be improved, not by contemplation of abstract ideas, but only by experience ... The bare contemplation of abstract ideas will carry us but a very little way in the search of truth and certainty. What, then, are we to do for the improvement of our knowledge in substantial things? Here we are to take a quite contrary course: the want of ideas of their real essences sends us from our own thoughts to the things themselves as they exist. Experience here must teach me what reason cannot: and it is by trying alone, that I can certainly know."[2]

Very early in the eighteenth century, Locke's essays gained enormous popularity among parents, educators, writers, and publishers. By the mid-eighteenth century a whole genre of children's literature dealing with child prodigies gained in popularity. Stories of wandering and wondering child prodigies, naively and preliminarily exposed to things in nature, presented children as closing the gap between man and nature and producing pure knowledge. Inspired by Locke's ideas, John Newbery (1713–67), an English publisher of children's books, adopted the motto *Deluctando monemus* (Instruction through delight). Among his titles was a widely popular series written under the pseudonym Tom Telescope, which went through seven editions between 1761 and 1787 alone. An introduction to Newtonian physics—the emerging science of the day—the series consisted of a group of lectures given by a precocious boy, Tom Telescope.[3]

Another seminal work on wild boys and children's connection to nature, *Émile* (1762), draws upon Locke's ideas. Jean-Jacques Rousseau presented his views about education in *Émile*, a semi-fictitious work detailing the growth of a young boy in the countryside, where, Rousseau believed, humans are most naturally suited, rather than in a city, where they learn only bad habits, both physical and intellectual. The first book of *Émile* starts off in this vein: "Under existing conditions a man left to himself from birth would be the most disfigured of all. Prejudice, authority, necessity, example—all the social conditions in which we find ourselves submerged—would stifle nature in him and put nothing in its place."

Later on, Rousseau makes the distinction between wild and civilized men by pointing to the image of the noble savage: "Natural man is everything for himself. He is the numerical unit, the absolute whole, accountable only to himself or to his own kind. Civil man is only a fractional unit dependent on the denominator, whose value is in his relationship with the whole, that is, the social body." In book 3, Rousseau outlines the educational ramifications of such a philosophy: "Make your child attentive to the phenomena of nature; soon you will make him curious. But to nurture his curiosity, never hasten to satisfy it. Put questions within his reach and let him solve them himself. Let him know nothing because you have told him, but because he has learnt it for himself. Let him not be taught science, let him invent it. If ever you substitute in his mind authority for reason, he will cease to reason; he will be a mere plaything of other people's opinion."[4]

Inspired by Rousseau, Thomas Day (1748–89) stressed the role of animals and nature in the raising of children. Day, a member of a circle of scientists, chemists, and inventors presided over by Erasmus Darwin, published *The Children's Miscellany*. He wrote a story called "Little Jack" (1788), about a young boy found by an old man. With barely enough grain for himself, the man did not know how he would be able to feed the child. After praying, however, he called his goat and was overjoyed to find that the boy suckled as naturally as if he had really found a mother.

Day's most famous production, *Sandford and Merton*, was a best seller for

Infant being baptized in a stormy river, inaugurating the process of education in nature. Frontispiece, Jean-Jacques Rousseau, *Émile, ou De l'éducation* (Amsterdam, 1762). Courtesy of Houghton Library, Harvard University.

A boy suckled by a goat, echoing Ḥayy Ibn-Yaqẓān's nourishment from the gazelle and from nature. Frontispiece, Thomas Day, *The Children's Miscellany* (London, 1797). Courtesy of Houghton Library, Harvard University.

eighty years. Published in three volumes (1783, 1786, and 1789), it tells of how rebellious Tommy Merton comes to see the error of his ways through the process of self-discovery. Made up of a host of linked stories, *Sandford and Merton* provides introductions to ancient history, astronomy, biology, science, exploration, and geography that enable facts and figures to be absorbed relatively painlessly through an engrossing narrative.

The paradigm of the wild boy appeared in other contexts, as well. Sarah Trimmer, for example, took the same trend in another direction, teaching children God's wonders through the exploration of nature. In the preface to her *Easy Introduction to the Knowledge of Nature* (1780), she writes that the book "[contains] a kind of general survey of the works of Nature that would be very useful, as a means to open the mind by gradual steps to the knowledge of the 'supreme being,' preparatory to their reading the holy scriptures."[5]

The works of Locke and Rousseau introduced not only a philosophy that based epistemology on the mechanistic relationship between sensations and reflections but also a new cultural icon of wisdom—the isolated wild boy who could directly build up knowledge from sensations and reflections on the blank slates of their own natures.[6] They facilitated the presentation of autodidacticism in the Enlightenment through diverse intellectual, scientific, and cultural notions—empiricism, trial and error, self-education of children and women, the return to the true natural self, prodigies as icons of wisdom, survivors cast on desert islands. Although an intertextual connection to the story of *Ḥayy Ibn-Yaqẓān* and to Ibn-Tufayl's experience-based philosophy can be detected, it is silent evidence that links the medieval utopia to the Enlightenment's multi-faced notions of autodidacticism, indicating a connection that goes further than mere intertextualities.

Several previous books had moved through diverse interpretive communities. *The Fables of Bidpai*, for instance, passed through culturally defined interpretations in the process of being translated from Sanskrit into various Near Eastern languages and then several European languages in medieval and early modern times.[7] Similarly, *Ḥayy Ibn-Yaqẓān* passed through several fields of cultural hermeneutics and, by the time of its translation into European languages during the early years of the Enlightenment, represented a source of inspiration for thinkers like Daniel Defoe, Locke, and Rousseau. But the text took a winding path to get there, moving across both time and space. Ibn-Tufayl originally wrote the piece, in Arabic, in Marrakesh in the 1160s. Between the twelfth and fourteenth centuries, the text moved north to the Iberian Peninsula, where it was translated into Hebrew. During the fourteenth and the fifteenth centuries, the Hebrew text traveled across the Pyrenees through Provence to Florence, where in 1492 Giovanni Pico della Mirandola had it translated for the first time into Latin. In the sixteenth and seventeenth centuries the story of the young boy on a desert island was in the background of those reading and producing utopian writings. Finally, in 1671 Edward Pococke produced a direct translation from the Arabic manuscript into Latin and titled it *Philosophus autodidactus*. A Dutch translation made by a mysterious S. D. B. appeared almost immediately in Amsterdam, in 1672, under the title *Het leven van Hai ebn Yokdan* (The life of Ḥayy Ibn-Yaqẓān).

In the wake of Pococke's publication, numerous translations from the Latin appeared in London; first in 1674 by George Keith, as *An account of the Oriental Philosophy*; again in 1686 by George Ashwell as *The History of Hai Eb'n Yockdan an Indian Prince, or The Self-Taught Philosopher;* finally, the most popular translation, in 1708 by Simon Ockley, under the title *The Improvement of Human Reason Exhibited in the Life of Hai Ebn Yokdan*. A Latin-to-German translation appeared in 1726 as *Der von sich selbst gelehrte Weltweise* (The self-taught world wise) by Georg Pritsius, in Frankfurt, and again in 1783, as *Der Naturmench* (The natural man) by J. G. Eichhorn, in Berlin.

Ḥayy Ibn-Yaqẓān in its many forms generated an enormous amount of interest on the part of contemporary philosophers and writers, especially Pococke's translation. For example, in 1930 Antonio Pastor, in his book, *The Idea of Robinson Crusoe*, noted the analogous connection between the plots of *Ḥayy Ibn-Yaqẓān* and *Robinson Crusoe*.[8] He presented circumstantial evidence that Daniel Defoe read the various English translations of *Ḥayy* and used them as a source of inspiration for his own story. Leibniz, too, seemed to know of *Ḥayy*. In a letter written to Albrecht von Holten (February 17–27, 1672), he expressed his admiration for Pococke's translation and displayed some familiarity with the figures of Ibn-Tufayl and his contemporary, Averroes.[9] Christian Huygens read copies of the Dutch translation, *Het leven van Hai ebn Yokdan* (1672), and one was found in Spinoza's library. Some even speculate that behind the translator's acronym S. B. D. stands Spinoza himself, who related to the identification of the oneness of God with nature.[10] Regardless of the truth of that rumor, the publisher of Spinoza's *Principles of Cartesian Philosophy* certainly did commission the Dutch translation.[11] The lack of a French translation is perhaps best explained by Voltaire's comment to a friend that while he wanted to translate the Latin translation into French, he decided not to, because the work is "spiritually nauseating romance."[12]

Locke, however, weighed the text seriously. As demonstrated in chapter 4 of this volume, Locke read Pococke's *Philosophus autodidactus* before writing his essays

Ḥayy's first experiment, dissecting the dead body of his mother-gazelle. The discovery of the secrets of nature represents a moment of enlightenment and communion with universal reason. Frontispiece, Dutch edition, *Het leven van Hai ebn Yokdan* (Amsterdam: William Lamsveld, 1701). Courtesy of Amsterdam University Library.

Left, illustration of the experimental practices of Ḥayy as he dissects his mother-gazelle; right, title page of the Ockley translation. Simon Ockley, *The Improvement of Human Reason Exhibited in the Life of Hai Ebn Yokdhan* (London, 1708). Courtesy of Houghton Library, Harvard University.

on human understanding and on education, directing following generations of writers to link the *noble savage* with self-learning—Rousseau, for example, who turned natural education into a systematic pedagogy of autodidacticism.

The challenge lies in recapturing the connections between *Ḥayy Ibn-Yaqẓān* and European writers and texts through five centuries and across diverse cultures. This book aims to show that more than coincidence lay in the treatise's presence at particular historical moments. Contemporary cultural controversies induced custodians and translators, who often identified with *Ḥayy*, to translate and diffuse the treatise to provoke discussions on autodidacticism.

In our age of academic professionalism, investigations into the text have been conducted through specific, and discrete, areas of specialization, from the discipline of medieval philosophy, which mostly employed textual analysis, to the discipline of comparative literature, which searched for the "influences" of the text on various writers during the Enlightenment.[13] This volume recovers the historical lacuna between medieval philosophy and the Enlightenment and explores the circulation of the text through the perspectives of history of science and intellectual history, following the text as it traveled from one culture to another, from one historical epoch to the next. Once used as a thread, instead of as self-valued text, here Ḥayy leads to various historical moments and to the recovery of particular cultural contexts in which its custodians and readers participated in controversies over autodidacticism and experimentalism. In recovering the mechanisms by which the treatise circulated across cultures—starting with medieval Andalusia and moving all the way to early modern Europe—the current cross-cultural historical exposé brings the treatise into play as a thread to various historical processes that fostered early modern notions of tabula rasa, the child prodigy, and experimentalism.

Accounting for five centuries of circulation through vastly diverse cultural contexts is, however, a large project, especially if the goal is to be as historically and culturally specific as possible. In my previous book, *Cross-Cultural Scientific Exchanges in the Eastern Mediterranean, 1560–1660* (2010), I deal with a similar problem of contextualization in cross-cultural historiography: how to settle the seeming contradictions between microhistory and cross-cultural history, between airless accounts of practices and airy accounts of the flow of ideas. I offered to follow scientific objects through different cultural contexts, recovering the mechanism of their motion and enabling the various cultural contexts to be revealed. The history of *Ḥayy Ibn-Yaqẓān*, however, goes on for far too long for a merely microhistorical approach to be useful. Instead, looking at the treatise from diverse cultural angles and at discrete moments in its circuitous history provides yet another mode to build up a widely prismatic and deeply contextual account. These moments pass through Marrakesh of the twelfth century, when Sufi challenges compelled Ibn-Tufayl to write his revolutionary philosophical treatise; to 1348 Barcelona, when pedagogical controversies engendered a commentary on its Hebrew translation; to late fifteenth-century Florence, where anti-astrology sentiment drove the translation into Latin; and finally to Oxford in the late seventeenth century, where the treatise embodied the contemporary experimentalist and empiricist visions. At the center of my investigations lies not the sheer reception of the story but rather what "was there" in these historical moments that compelled custodians and readers of the text to refer to it and translate it.

Isolated historical moments hinder the effort to give a coherent history of the treatise; seemingly only its generic message of self-teaching binds them together.

The organization of the book, therefore, ought to bring together what, on the surface, is a collocation of empirically grounded local meanings into a flowing cross-cultural narrative. The art of sampling in contemporary music and in visual art offers a method to bring together freestanding objects, sequencing them in a new order so that new meaning arises out of their conjunction. The film *The Red Violin*, for instance, tells various stories of a fine musical instrument that passes from generation to generation of classical virtuosos, with each new cultural milieu in a sense adapting the instrument to its own forms of production. The appropriation of such methodological principles yields *historical sampling*, to coin a phrase, an organizing framework for cross-cultural accounts that stand as both microhistorical and prismatic. Historical sampling allows a unique perspective on the diverse cultural perceptions and interpretations of the quintessential treatise on autodidacticism—*Ḥayy Ibn-Yaqẓān*.

Local meanings, however, come to light neither through text analysis nor in comparing the translations to the original text nor even in focusing merely on its direct reception. An exploration of adjacent historical phenomena and a focus on their proximate relations to particular textual objects that carried the story of *Ḥayy* morphologically puts together the pieces of the puzzle into a coherent picture, recovering webs of local meanings of autodidacticism. Sources, current events, political movements, and human networks once arranged around and in proximity to the text bring in thick descriptions, unfolding historical complexities that elevated autodidacticism as the prime program for knowledge acquisition.

CHAPTER ONE

Taming the Mystic
Marrakesh, 1160s

WHILE IN ALEPPO IN THE 1630S, serving as an official in the Levant Company, Edward Pococke purchased a manuscript of *Ḥayy Ibn-Yaqẓān* and used it for his Latin translation *Philosophus autodidactus*, gifting it later to the Bodleian Library, where it still resides.[1] Interestingly, the manuscript actually bound together the story of *Ḥayy* with *Al-Maʿārif al-ʿaqlīyah*, a work of the mystic-philosopher Abū Ḥāmid al-Ghazzālī (known in Latin as Algazel), implying the connection of the mystical writings of al-Ghazzālī to the philosophical tale of Ibn-Tufayl.[2] The implication is not without grounds. From the twelfth through seventeenth centuries, manuscripts of *Ḥayy Ibn-Yaqẓān* circulated throughout the Near East, promoting a program that favored firsthand practical experience over traditional textual authority. Moreover, the arrangement of political movements, human networks, and adjacent texts around Ibn-Tufayl, and the examination of their proximate relations with *Ḥayy Ibn-Yaqẓān*, brings to light a convoluted dialogue Ibn-Tufayl actually carried on with his mystical surroundings.

From the beginning of the twelfth century, al-Ghazzālī's works circulated westward to Morocco and Spain, when Sufism had started to gain a strong hold in Andalusia, challenging traditional forms of authority. The Sufi self-discovery of God through physical rituals not only challenged political authorities but also subverted the rationalist tenets of Andalusian philosophy, prompting Ibn-Tufayl to write his text.[3] The Islamic discovery of the southern Indian Ocean generated tales about fabulous utopian islands, providing Ibn-Tufayl the place and the culture for his thought experiment. His aim was to tame the mystic practice of self-discovery by subjecting it to philosophical procedures. The result was a presentation of autodidacticism as a philosophical program that strikes a balance between senses and reason, textual authority and oral culture, and generally, between theory and practice.

Abū Bakr Muhammad Ibn-Tufayl (in Latin, Abubachar) was born in Wādī Ash (present-day Guadix), a small town of houses built into the caves northeast of Granada, in 1105 and died in al-Maghrib (present-day Morocco) in 1185. Born into social turmoil, he grew up in the midst of political transition. For a century (1040–1147), the Almoravid dynasty had built unified rule over northwestern Africa and Muslim Spain. Between 1130 and 1145, while northern Christian military campaigns forcibly wrested Saragossa and other central Iberian cities from them, the Almoravids faced a severe political challenge from within. Enthusiastic Sufis challenged the traditional form of authority and its political base—the alliance between the jurists and the rulers. In the Atlas Mountains of al-Maghrib, a coalition of Berber tribes under the leadership of Muhammad Ibn-Tūmart (d. 1130) started pressing from the south and cracked open the Almoravid empire, finally taking over their capital, Marrakesh, in 1147. The succeeding Almohad dynasty gradually extended its power over northern Africa and later over al-Andalus, eventually moving their capital to Seville in 1170. Ibn-Tufayl served the young Almohad dynasty, first as a secretary to the ruler of Granada and later, during the 1160s, in the court of the young sultan Abū Ya'qūb Yūsuf (1163–84), as a minister and as a court philosopher. In these capacities, he helped his ruler cope with political and theological challenges that threatened political stability.

One day, during the 1160s, probably after a full day of meetings with various officials about the political unrest caused by continuing Sufi revolts, Ibn-Tufayl closed himself in his chamber and sat down to write a philosophical tale about a little boy stranded on a desert island. In the back of his mind traumatic political and religious events that had taken place fifty years earlier must still have echoed.

In 1109 the notables of the city of Cordoba gathered by the western gate of the city's main mosque to witness the burning of al-Ghazzālī's voluminous masterpiece, *The Revival of the Religious Sciences* (*Ihyā' 'ulūm al-dīn*). The work had been condemned by the chief judge of Cordoba, Husayn Ibn-Hamdīn (d. 1114) and other jurists of the city and ordered destroyed by the Almoravid sultan, 'Ali b. Yūsuf b. Tāshufīn (d. 1143). The sultan's representatives collected all available manuscripts, and many who had come under suspicion of possessing a copy had to swear that they did not. The leather-bound volumes were doused in oil and ignited.[4]

Still, rulers burn books not so much for their words and ideas as for the readers they attract. The Almoravid dynasty, which relied heavily on jurisprudence, viewed *The Revival* as an imminent threat to social and political order. Most studies of the burning portray *The Revival* as a work of Sufism that criticized

Map of al-Andalus and al-Maghrib. Although the Muslims lost control of the northern parts of the Iberian peninsula, for some centuries al-Andalus and al-Maghrib were administered as one political and cultural unit.

the role of jurists and theologians in the practice of Islam. Reinhard Dozy, the pioneering historian of al-Andalus, describes *The Revival* as representing "an intimate, fervent, passionate religion, a religion of the heart." He notes that al-Ghazzālī blamed the religious scholars of his day for occupying themselves with trivial matters of law. The jurists of al-Andalus, whose "intolerance passed all bounds," Dozy stresses, became incensed at this critique of their practice and ordered the work burned.[5] Kenneth Garden has forcefully argued that in *The Revival*, "al-Ghazzālī sought to revive Islamic religious sciences by subordinating the 'worldly science' of jurisprudence (*fiqh*) and theology (*kalām*) to the 'otherworldly science,' a science largely, but not entirely, identical to Sufism."[6]

The Revival, indeed, does represent a polemic, and as its title suggests, it aims to revive the religious tradition—a tradition distorted by the jurists. Al-Ghazzālī accused the jurists of wrongly reducing the science of religion (*'ilm al-dīn*) to jurisprudential discussions that employed jurisprudential language, which excluded other valuable intellectual fields. Moreover, ordinary men, either illiterate or not fluent in the legal jargon, could not participate in intellectual life at all because the inaccessibility of legal language kept intellectual discussions and interpretations within the confines of small jurisprudential (*fuqahā'*) circles.

Al-Ghazzālī believed that the terminology of jurisprudence had become unclear because the scholars using it kept themselves at a remove from the pure natural origins of wisdom that had been laid out centuries before. As a result, he traced a historical timeline of the development of knowledge, portraying it as a fall from grace, a decline from pure knowledge into words, characterizing the early stages of Islam as possessing an otherworldly and oral knowledge. Only in the second and third centuries of Islam did this knowledge take written form, and then the jurists (*fuqahā'*) made a science (*'ilm*) of it that amounted to no more than giving legal rulings. The true fall from the pristine state of knowledge came in the tenth century, when books of theology (*kalām*) began to gain legitimacy and the practice of debating points of theology came into being. The otherworldly sciences were forgotten by the common people, while theologians and jurists rose to the heights of intellectual and religious authority.[7]

To cover this historical lacuna, al-Ghazzālī outlines an autodidactic theory of practice. Practical exploration of nature and God, he accentuates, recaptures the forgotten science of religion and leads the practitioner, in the final phase, to ecstasy. He lays out a trajectory for the practical and gradual acquisition of knowledge, from the most inferior to the most exalted divine knowledge. Legal rituals represent the first step of the practitioner's path of self-discovery, followed by cleansing of the heart and ego, recitation, and meditation, finally achieving communion with God. This ladder outlines how the "science of practice" (*'ilm al-mu'āmala*)—based on firsthand experiences in the physical world—can rise to the level of the "science of unveiling" (*'ilm al-mukāshafa*), wherein divine secrets of nature and God are revealed.[8]

Al-Ghazzālī's fusion of theory and practice has theological ramifications. Since he prefers action to language, a limited tool in his view, he sees textual authorities as having lost their power over the private experience of the individual.[9] But the loss of traditional authority can lead to a lessening of trust in any knowledge based in firsthand experience. Al-Ghazzālī posits the sincerity (*al-ṣidq*) of the practitioner engaged in the search for knowledge as the sign of credibility. The process itself leads the practitioner to become keen sighted (*baṣīr*), and this very awareness puts him in opposition to jurists and theologians, who cynically use the religious sciences for political power, financial gain, and self-interest.[10]

The Almoravid dynasty and its allied jurists, then, had many reasons to worry about the potential of *The Revival* to inspire social disorder. The book—and its author—disregarded the jurists' authority and offered unguided self-discovery as the path to divine truth. It encouraged the acquisition of knowledge through firsthand experience over that of transmitted tradition. And finally, in its disdain for textual tradition and its focus on practice and oral tradition, *The Revival* gave everyone, even illiterates, the possibility of experiencing God. The closed circle of jurists, and their prerogative to exclusively interpret the written law, now lay in danger. Thus in 1109 they took action. The leading jurist, Ibn-Ḥamdīn, launched a campaign to collect manuscripts of *The Revival* and burn them.

Mocking the Sufis

During those years, chief judge Ibn-Ḥamdīn wrote his refutation of *The Revival*. Only a fragment of the refutation has survived; in it, Ibn-Ḥamdīn expresses his concern about the social and political implications of a growing Sufi group called Ghazzalians, who drew nourishment from *The Revival*. In targeting this Sufi group, Ibn-Ḥamdīn ridicules the practice of embodied contemplation (*dhikr*), giving the first historical evidence for its existence in al-Andalus. In this practice, the Sufi moves in circles and on his own axis, focusing on the remembrance and repetition of the names of God. Only when the practitioner can aver, "I am God" and "I am the Truth" can full communion through the practice be achieved.

Ibn-Ḥamdīn's main objection to *The Revival* focuses not so much on practice as on the apparent Sufi disregard for traditional textual authority. In particular, he rejects al-Ghazzālī's attempt to supersede tradition as a whole and resents the lack of effort to rationally explain how firsthand experience of God and nature could lead to knowledge. He charges that Sufis lacked the training needed to address the question of the nature of the divinity, and he ridicules their autodidactic path to knowledge through the practice of *dhikr*: "Sufis learn the divine sciences without studying." Ibn-Ḥamdīn stresses, "The Sufi sits with his heart empty, concentrates and perpetually recites 'Allāh, Allāh, Allāh' . . . and says that 'when one obtains this climax he must seek isolation in a dark room and be wrapped in his cloak and at this time he will hear the address of the Truth: '*Oh thou enveloped in thy cloak [al-mudaththaru]!*' and '*Oh thou wrapped up in thy raiment [al-muzammalu]!*'"[11]

The struggle went beyond the limits of distant intellectual discussion and

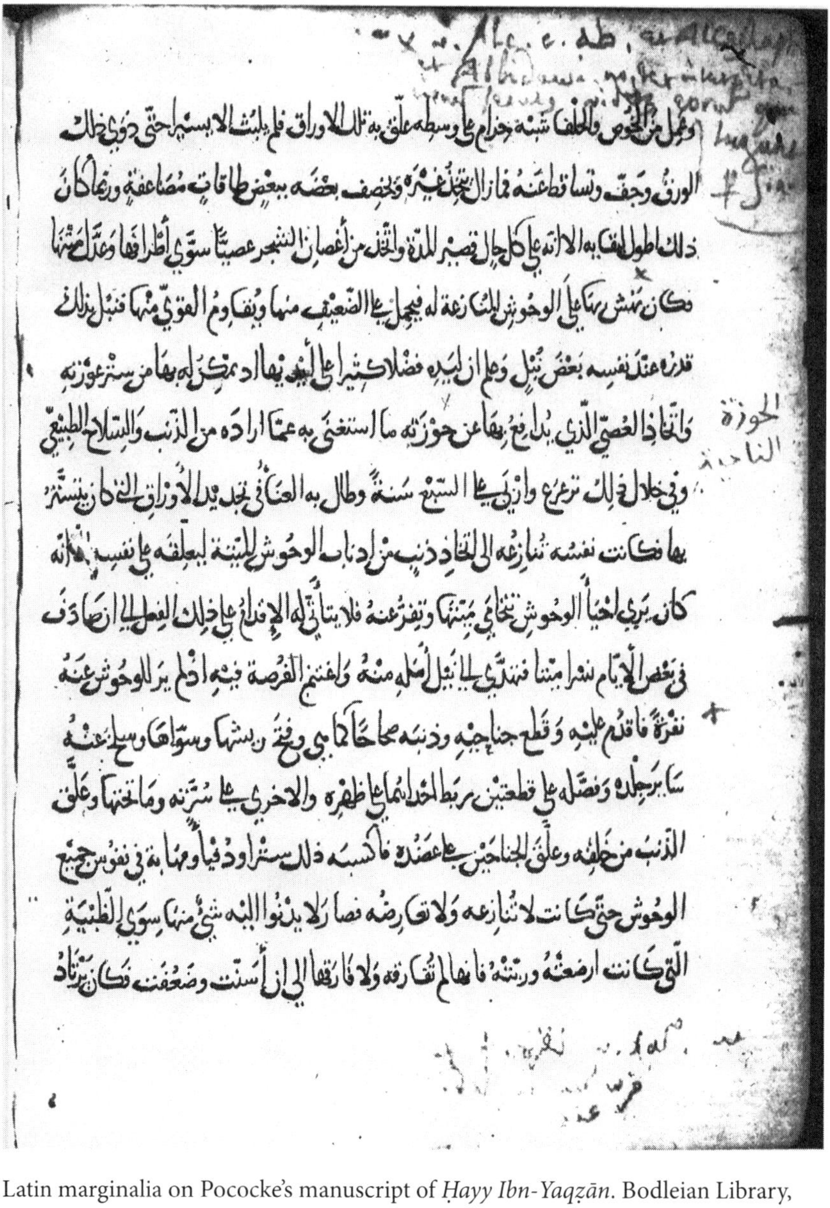

Latin marginalia on Pococke's manuscript of *Ḥayy Ibn-Yaqẓān*. Bodleian Library, MS Pococke 263. Courtesy of the Bodleian Library, Oxford University.

writings and involved personal rivalry. The chief judge of Cordoba, Ibn-Ḥamdīn, a leading theological authority, attracted students from different parts of North Africa, one of whom, Muhammad Ibn-Tūmart, became the founding father of the contending Almohad dynasty.

Ibn-Tūmart left his Atlas Mountain home and arrived in Cordoba in 1106, three years before the burnings of *The Revival*. He spent a year in Cordoba and, as the fourteenth-century historian Ibn Qunfudh indicates, "studied with the chief judge, Ibn-Ḥamdīn, and later on embarked at Almeria on a boat that headed to the East, where he" allegedly "studied with al-Ghazzālī."[12]

Such myths about personal connection between the key figures further propagated relics of the controversy to oral and popular culture for the next generation. Fifty years later, after the Almoravid dynasty had fallen, the Sufi unrest continued, and Ibn-Tufayl, courtier of the new Almohad dynasty, entered into dialogue with *The Revival*, which, despite the oaths taken, had left vestiges in society and culture.

Ibn-Tufayl not only was aware of the heated controversy over the reception of *The Revival* but also rather closely read Ibn-Ḥamdīn's critique. Fifty years later in the opening paragraph of *Ḥayy Ibn-Yaqẓān*, he scorns the Sufis for their pretentious claim that they could achieve divine ecstasy without recourse to gradual, rational exploration of nature and God, reiterating the words of the Cordoban rationalist judge and ironically parroting the Sufis ecstatic cry, "*I am the truth*."[13] The relations between *Ḥayy Ibn-Yaqẓān*, al-Ghazzālī's mystical *Revival*, and Ibn-Ḥamdīn's rationalist *Refutation* indicate that Ibn-Tufayl was not only closely reading the two earlier texts that became pivotal to competing political forces but had also turned them into the prime materials for his settlement between the incoming Sufism and the rooted rationalist tradition of al-Andalus.

In the introductory paragraph, addressed to a student, Ibn-Tufayl appropriates Sufi jargon while at the same time rejecting Sufi claims for direct knowledge of God. He tells his student, who seems to have been so immersed in mysticism that his plea to disclose the secrets of oriental philosophy "evoked in me a noble idea and has brought me to a direct sensible intuition (*mushāhadah*), a state which I have not experienced before, and has promoted me to the furthest level, that a tongue cannot describe." Ibn-Tufayl used the notion of direct sensible intuition, which was heavily used by Sufis, to stress the need for an exploration of nature through the senses. The preference for action over words accentuates the thought that the language of ideas cannot really get at the secrets of nature, which can be revealed only through firsthand experience. Right from the start, then, Ibn-Tufayl states that knowledge and experience go hand in hand. The secrets of oriental philosophy, which he claimed to deliver, actually constituted an autodidactic program to explore nature by means of direct intuition and firsthand sensations. Ibn-Tufayl, however, stresses that his notion of sensible direct intuition does not resemble that of al-Ghazzālī's Sufi followers and maintains that arguments for a sensible experience can be misappropriated if used without scientific principles and foundations, especially in the case of the Sufi who "is ignorant in the Sciences, and yet falsely claims experiencing the ultimate truth."[14]

According to Ibn-Tufayl, these pronouncements must, by definition, be insincere, because visions of truth and God cannot be described or transmitted in words. To further subvert the claims of the Sufis he brings up the words of their master al-Ghazzālī, who wrote in *Al-Maʿārif al-ʿaqlīyah* (bound together with *Ḥayy Ibn-Yaqẓān*) that when one reaches this state of ecstasy of pure vision of God and the Truth, one should not speak about it, since it is not expressible in words.[15]

The controversy over *The Revival* thus represented the silent backdrop for the writing of *Ḥayy Ibn-Yaqẓān*. In reacting to it, Ibn-Tufayl merges al-Ghazzālī's science of practice with the philosophical tradition of al-Andalus that emphasizes logic, mathematics, and astronomy. Standing on the shoulders of the mystic theologian al-Ghazzālī and the rationalist Cordoban judge, Ibn-Ḥamdīn, Ibn-Tufayl takes their arguments a step further by philosophically taming the Sufi science of practice. On one hand, he accepted the Sufi argument that the secrets of nature are attainable while still rejecting the claim that ecstasy is possible without gradual philosophical practice. On the other, he accepted Ibn-Ḥamdīn's challenge and tried to give to al-Ghazzālī's science of practice a rational underpinning.

There is also an allegorical connection to al-Ghazzālī's *Revival*. Ibn-Tufayl's use of the parable of the nursing gazelle (rather than of a goat, a wolf, or any other mammal) encodes al-Ghazzālī's name ("gazelle") into the very fabric of his treatise. Ḥayy takes nourishment from al-Ghazzālī as a starting point from which he tries to synthesize the logical and mathematical tradition of the Maghrib with Eastern theories of knowledge and mysticism. He cites one of al-Ghazzālī's verses describing the limits of philosophical language, "whereof one cannot speak, thereof one must be silent," to emphasize the notion that one should not confine philosophy to words but should instead explore, firsthand, in order to attain knowledge of the truth. Ibn-Tufayl, however, turns this philosophy of language on its head. Whereas al-Ghazzālī's citation implies that ultimate mystical experience cannot be described by or understood through the sciences, Ibn-Tufayl counters that full knowledge of nature and God can not be truly attained without philosophical procedures guided by reason nor without a gradual exploration of nature that corresponds with the philosophical hierarchy of disciplines.

Ḥayy's Ladder of Philosophy

In *Ḥayy Ibn-Yaqẓān*, then, Ibn-Tufayl blended Andalusian polemics over Sufism with Eastern cultures and practices to show that the fundamental principles of physics, metaphysics, and theology can be revealed without teachers. Through sensible perception Ḥayy explores the natural world and teases out its secrets by trial and error, inferring its fundamental laws. The story climaxes with Ḥayy's search for a direct sensible intuitive perception that would elevate him to the highest level of beingness. His observations of the stars lead him to conclude that they represent the closest tangible things to God. He outlines three attributes of the stars—sublunar, celestial, and metaphysical—and begins an imitation of them that eventually leads him to communion with the "necessary existent being," God.[16]

It is here that Ibn-Tufayl takes a step toward synthesizing a philosophical theory of knowledge with Sufi practice, a synthesis that necessarily considers man the superior creature and nature a passive organ awaiting an active force to conquer and control it. The practical aspect of natural philosophy begins with assimilating a sublunar attribute—the preservation of nature and ecology. His first step, resembling the sublunar world, the world of generation and decay, aimed at taking over nature by "removing from animal or plant any harm or damage, or impediment: and when he did cast his eyes upon any plant . . . he would remove whatever was interposed . . . He would water plants, so far as he could; and when he would look for an animal which some predatory beast was pursuing . . . or that hunger and thirst had seized it, all these things he did undertake to remove with all his power."[17]

Ḥayy's assimilation of the celestial world, eternal and pure, represented a kind of bodily practice guided by hygienic rules. Imitating the infallible stars, shiny and clean, Ḥayy "kept himself in a continual cleanliness, by removing all impurity and filthiness from his body, and by washing himself with water, and purging his nails and his teeth, and also the secret parts of his body . . . until they had splendor, beauty, and cleanness, until he wholly shined." In imitation of the celestial sphere's perfect circular motions, Ḥayy "used diverse kinds of circular motions, sometimes going around the island, its shore and utmost parts. Sometimes he compassed his house, or some rock with various circuits, either walking or running, and sometimes revolving around himself, until dizziness took hold of him." Finally, when Ḥayy attempted to focus on the necessary existent being in order to assimilate the metaphysical attribute, he removed "from himself all impediments of sensible things, and shut his eyes, and stopped his ears, and by all his strength he restrained himself from following his imagination, endeavoring as much as he could to mind nothing but *himself* . . . by revolving himself about, until all sensible things vanished . . . until his cogitation would be pure from mixture, and thereby, promoting him to the next stage to perceive that *necessary existent being*."[18]

Through these practices, Ḥayy spiritually travels to the otherworld and experiences ecstatic communion with God: "He confined himself, in the lowest part of a cave, where he sat quietly, his head bowed down, his eyes kept low, and he turned away from himself and from all sensible things and bodily faculties, he bent his mind and thoughts upon this one necessary existent, and did not admit any other thing . . . When then he was deeply plunged in the vision of *that first being, the true necessary existent* . . . he saw what neither the eye has seen, nor the ear heard, nor came into the heart of man to conceive it."[19]

In book 7 of Plato's *Republic,* where the metaphor of the cave appears, the lowest part of the cave represents the realm furthest from philosophical knowledge. Darkness, shadows, and opinions reside here, in opposition to the realm of enlightenment and knowledge. In contrast, Ḥayy's cave is the place where one can immerse one's self in pristine knowledge.

In his quest, then, Ḥayy strives for the ultimate purpose of any medieval search for wisdom, communion with God; but as the story demonstrates, even that should be explored through practice. Indeed, the story enacts Ibn-Tufayl's philosophical theory of acquiring knowledge in such a way that its climax becomes like a Sufi ritual. Unlike the Sufis, however, whom Ibn-Tufayl mocked right from the outset, Ḥayy achieves communion with God not through mystical rituals but through gradual philosophical practice. "For there is no way of finding out what truly occurs at this life of experience," Ibn-Tufayl concludes, "besides reaching out for it." Thus he makes Ḥayy climb the ladder of philosophical disciplines: trial and error in physics, heavenly observations, and finally, metaphysical deductions through asceticism, circular dance, and meditation. Experience was Ibn-Tufayl's alternative to the traditional philosophical contemplation of nature and God.

Ibn-Tufayl, then, attempted to tame the Sufi theory of practice and to subject it to philosophical procedures and disciplinary hierarchy. The question left unanswered, though, is why the controversy over Sufism, and the fiercely held positions that animated it, should still be pertinent fifty years after it broke out, when Ibn-Tufayl sat down to write Ḥayy Ibn-Yaqẓān.

The particular cultural circumstances are hard to discern, because Ibn-Tufayl left little information about his life. Only two texts make reference to him at all: Ḥayy Ibn-Yaqẓān, which he wrote, and two paragraphs in *The History of the Almohads* by 'Abd al-Wāhid al-Marrākushī, a thirteenth-century historian who notes that Ibn-Tufayl held a ministerial position at the court of Abū Ya'qūb Yūsuf, the third Almohad sultan. Nor, for that matter, is there an accurate date for the completion of Ḥayy Ibn-Yaqẓān (I suspect that it was composed around the 1160s). Proximate historical phenomena, however, point to the cultural context to which he responded.

The Sufi Revolt and the Almohad Court

The 1109 burning of *The Revival* did not put an end to its reception and circulation. During the 1130s and 1140s, when political rule shifted from the Almoravid to the Almohad dynasties, Sufism emerged as a major social and political force, with *The Revival* as its authoritative guide.[20] Groups of Maghrib Sufis appeared toward the mid-twelfth century in response to newly heightened tensions between ascetically minded religious scholars who were suspicious of political authorities and jurists who benefited from their increased status under the Almoravids. Responding to the system in which jurists were paid directly from the state treasury, and so were fully integrated into the state apparatus, Sufi scholars objected to the symbiosis between religious and political authority and posed their own asceticism and firsthand experience as an alternative.[21] The 1109 burning took on a significance the Almoravids did not foresee, and *The Revival* become a symbol of Sufism itself.[22] By the mid-twelfth century, the Almohads, who were on the verge of toppling the Almoravids, circulated legends of their own, linking their founder, Ibn-Tūmart, to al-Ghazzālī and noting his objections to the earlier burning of *The Revival*. Thus the Almoravids viewed Sufis and Almohads as political and religious allies who threatened the existing political order, leading the Almoravids to persecute Sufis. As a result, a Sufi revolt began in 1144 under the leadership of the charismatic Abū al-Qāsim Ibn Qasī.[23]

Two incidents sparked the flames of revolt. In 1141 Ibn al-'Arif, the leader of a group of Sufis known as the Ghazālian order (*al-ṭarīqa al-ghazāliyya*), and Ibn Barrajān, known as the "al-Ghazzālī of al-Andalus," were both summoned to the Almoravid capital of Marrakesh for questioning. Ibn Barrajān was executed on the spot, and Ibn al-'Arif died on his return to al-Andalus under disputed circumstances. In the southwest Iberian peninsula, a self-styled Sufi messiah named Ibn Qasī is said to have defied the government by publicly reading al-Ghazzālī's *Revival* and calling others to arms. Eventually captured and brought in chains to the Almoravid capital, Marrakesh, he too was executed, immediately attaining martyrdom in the eyes of the Almohad movement.[24] Finally in 1144, three decades after the first burning of *The Revival*, the Almoravid sultan Tāshufīn Ibn 'Alī (d. 1145) sent a statute to the people of Valencia ordering them to seek out and burn the works of al-Ghazzālī.

This later burning of *The Revival* embodied not just an intellectual struggle but a full-blown political clash. The chronologist Ibn al-Qaṭṭān, author of the thirteenth-century history of the Maghrib (*Nuẓum al-jumān*), says that the burnings of *The Revival* led to the overthrow of the Almoravid dynasty by the Almohads, whose founder had been "personally entrusted by al-Ghazzālī to avenge the destruction of his masterpiece."[25] This political turmoil, which was further extended to the 1160s, served as the backdrop of Ibn-Tufayl's activity as court philosopher for the Almohad sultan. He embarked on composing the perfect tool to advance the Almohad political theology and propaganda, a textual tool that would be used to deal with similar political and cultural threats.

The Almohad ideology, constructed by its Berber founding father, Muhammad Ibn-Tūmart, presented a challenging political theology of God's oneness (*tawḥīd*), which originated in arguments about hermeneutics. The previous dynasty, the Almoravids, promoted the Mālikī school of Islamic law, which was characterized by a rigid formalism and a strict adherence to the dictates of the Quran but led to some unsatisfactorily literal interpretations of problematic verses. Ultimately, jurists began to neglect scholarship of the Quran and the stories of the prophetic tradition (Ḥadīth) and focus instead on jurisprudence (*fiqh*). Al-Andalusian schools taught this rationalist view of religious studies, and the scholarship of philosophy that arose out of it carried its unique stamp in which logic, mathematics, and astronomy took center stage.

Once the Almohads and Sufis adopted al-Ghazzālī's *Revival*, they began to attack the jurisprudential outlook of their predecessors. Continuing al-Ghazzālī's critique of jurisprudence and theology, Ibn-Tūmart sought to establish Islam on its fundamentals—the Quran and Ḥadīth. In his main work, *The Most Precious Thing Asked For* (*A'azz mā yuṭlab*), he claims to have debated the jurists over the sources of truth and states that the profession of faith in God's Oneness (*shahāda*) should not inaugurate the acceptance of Islamic religious life but rather should come at the climax of gradual education, guided by reason.[26] For Ibn-Tūmart, the notion of God's oneness already exists in the believer's mind. Such knowledge could be invoked through exploration and research,[27] including the study of nature manifested in the widespread ecstatic rituals of the Sufis.

But Ibn-Tūmart's doctrine went beyond pantheistic monism and involved cultural values. Vincent Cornell has argued that Ibn-Tūmart's concern lay in the realm of individual responsibility.[28] The key to the philosophical and ethical Almohadi system can be summed up in Ibn-Tūmart's dictum, "Understanding is the mother of ability" (*al-Idrāk umm al-Istiṭāʿa*).[29] Thus professions of faith do not go far enough. Knowledge of the whole entails understanding its parts, which requires personal responsibility.

To popularize his vision, Ibn-Tūmart also demanded accessibility. He declared that any discussion of knowledge should be written so that broad circles of illiterates and Berbers would be able to understand the stories of the Quran and Ḥadīth.[30] In fact, his first act after taking control subverted the textual authority of Arabic books by writing a book on his vision of God's oneness for the Berbers "in the Berber language."[31]

Intellectual culture became more accessible. Scholars moved from using the philosophical language of logic and law to rendering ideas through narrative. The accessibility and popularization of knowledge, performed in public speeches, also facilitated the rendering of the dynasties' political theology to illiterates and thereby efficiently enabled the centralization of political power. The shift in the philosophical style quickly produced cultural and political capital, and Berbers and Sufis started aligning themselves with the Almohads against the Almoravids.

Almohadi bureaucrats, among them Ibn-Tufyal, began disseminating Ibn-Tūmart's new philosophical outlook.[32] The thirteenth-century historian ʿAbd al-Wāhid al-Marrākushī, states that Sultan Abū Yaʿqūb Yūsuf enabled the rich intellectual activity that characterized his reign; the sultan "continually gathered books from all corners of Spain and North Africa and sought out knowledgeable men, especially thinkers, until he had gathered more than any previous king in the West." Al-Marrākushī also provides additional details about Ibn-Tufayl's work: "I have seen works of Ibn-Tufayl's on both natural and metaphysical philosophy, to name only two fields of his philosophical competence. One of his books on natural philosophy is called *Ḥayy Ibn-Yaqẓān*. Its object is to explain the meaning of human existence according to philosophical ideas. The book, written in the form of letter, is slim but of tremendous benefit in this study."[33]

Al-Marrākushī gives some details about Ibn-Tufayl's everyday practices and illustrates the types of people with whom he interacted on a regular basis:

He hired many kinds of employees, among whom [were] physicians, engineers, secretaries, poets, archers, soldiers, and others ... Sultan Abū Ya'qūb Yūsuf loved him ardently; I learned that he would stay with him at his palace whole days, at night and daytime, not appearing [elsewhere] ... Ibn-Ṭufayl kept bringing to him scholars from all regions, directing his attention to them, inducing him to honor and acclaim them. He is the one who directed his attention to Ibn-Rushd (Averroes), and it was since then that he became known and his standing noted among them.[34]

The court philosophers of Sultan Abū Ya'qūb Yūsuf, apparently, produced the foremost philosophical texts combating Sufism, al-Ghazzālī, the *Hayy Ibn-Yaqẓān* of Ibn-Tufayl, and, later on, Averroes' *The Collapse of Philosophers*.

As Al-Marrākushī's excerpts indicate, Ibn-Tufayl mostly interacted with the sultan and with his official colleagues. Yet though *Hayy Ibn-Yaqẓān* takes the form of a story, its style and language were incomprehensible to the masses of illiterates and Berbers. Neither the masses nor the philosophers around Ibn-Tufayl made up his desired audience. It remains unclear for whom he wrote this allegorical telling of a philosophical system in the form of a story, leaving only the possibility that the intended audience for *Hayy Ibn-Yaqẓān* were the professional literati, for whom the book signified not a form of entertainment but a part of their education. *Hayy Ibn-Yaqẓān*, therefore, seems to lie not within a philosophical textual tradition but within the tradition of *adab* [belles lettres], which officials used to learn good manners, culture, and a broad and general control of knowledge. In *Hayy Ibn-Yaqẓān*, Ibn-Tufayl furnished officials in court with the basic foundations of philosophy, in accordance with Almohadi political theology, and stressed God's oneness in concord with the exploration of nature.

In the introduction, Ibn-Tufayl places his treatise within the history of philosophy in Andalusia, stating that early Andalusian philosophy focused on mathematics and, later, on the propagation of the philosophy of logic and theories of knowledge to the East. Some Andalusian philosophers achieved a little understanding in logic, but it did not bring them to true perfection. In *Hayy Ibn-Yaqẓān*, then, Ibn-Tufayl offered the professional literati a tool with which to reconcile an older al-Andalusian philosophical tradition with Almohadi political theology. He cast Hayy as the ultimate pious Almohadi who lives a completely ascetic life, devoted to experiencing the ecstasy of knowing God. Ibn-Tufayl took the framework set out by Ibn-Tūmart, making Hayy a responsible boy-man who explores nature for religious ends, guided by his rational ability. He turns *Hayy Ibn-Yaqẓān* into something like a guide for the perplexed for those who have not yet internalized the doctrine of divine unity. Ḥayy Ibn-Yaqẓān thus helped solidify the political theology of God's oneness and centralize the political rule of the Almohads. But solidification of ideology and centralization of politics entail intolerance.

While Ibn-Tufayl served in court as a high minister in the years leading up to and including the 1160s, political disturbances accumulated: Sufi orders threatened to rise up, the Muslims of Toledo faced persecution by neighboring king Alfonso VII of Castile, and a revolt in Granada occurred in which local Jews took part. In this new unstable political atmosphere, intolerance grew. Muslims who still held anthropomorphist views in their religious outlook were considered infidels and had to openly swear to their belief (shahādah) in the oneness of God in front of state officials. Infidels' property was confiscated and given over to the religious endowment. They were subject to special taxes (kharāj)—some of the most important sources of income for the state. Religious minorities were forced to wear special distinctive clothing (giar) and were forbidden to practice medicine. In some places they had to face the choice of conversion or expulsion.

Reconciling the philosophical open-mindedness of Sultan Abū Ya'qūb Yūsuf with his intolerance toward other religions, ideas, and customs seems well nigh impossible, but as a reading of the text quickly shows, Ḥayy Ibn-Yaqẓān embodied these contradictions. On the one hand, it presented a vision of universal man; on the other, that man turned out to be an Almohadi zealot.

In the face of widespread persecution, philosophers were not always prepared to express their views on all philosophical subjects. Al-Marrākushī gives an account from a student of Averroes who overheard his teacher describe his introduction to Sultan Abū Ya'qūb Yūsuf. The subtext of Averroes' description points to the fear of persecution: "When I went in to the Sultan Abū Ya'qūb Yūsuf," Averroes relates, "I found him alone with Ibn-Tufayl. Ibn-Tufayl began praising me and speaking of my family and my background, very kindly adding many good things that I really did not deserve. Having inquired as to my name and origins, the first thing the sultan asked me was 'What do they [philosophers] believe about the heavens? Are they eternal or created?' I was seized with concern and did not know what to say. I tried to excuse myself by denying that I had studied philosophy. I had no idea how far his prior discussions with Ibn-Tufayl had gone."

Court philosophers had reason to be sensitive about persecution, but Averroes had to be particularly cautious in the Almohadi court. The reference to Averroes' family and ancestors was an uncomfortable reminder of his grandfather, one of the judges involved in the burning of al-Ghazzālī's Revival.[35] His fear in the face of the sultan's question on the eternity of the universe indicates that the borders of intellectual intolerance remained ill-defined, with the sultan seemingly allowing himself certain philosophical discussions that were not allowed outside of court.

Regardless of his fear of persecution, Averroes ends the story well enough:

> His Excellency saw that I was frightened and confused. He turned to Ibn-Tufayl and began to discuss the question with him, referring to the positions of Aristotle and Plato and all the other philosophers, [and] citing the arguments of the Muslims against them. I soon realized that he was more learned than I would have expected a full-time professional to be. He put me so well at ease that I myself spoke up and he soon saw that I was not as ignorant as I had seemed. When I had gone he sent me a gift of money, and a splendid robe of honor and a horse.[36]

In trying to appropriate Sufism, the Almohadi sultans acted on the recent memory of the Sufi uprising that eventually helped to topple the Almoravids. For Sufis bestowed authority based not on genealogy, efficient organization, or law but on charisma and other leadership qualities. The political and religious crises of the mid-twelfth century led to disillusionment with traditional forms of authority, and a new type of leader arose: the Sufi master.[37] The Almohadi rulers feared that Sufi leaders of the previous revolt might inspire others.[38] Ibn-Tufayl's patron, Sultan Abū Ya'qūb Yūsuf, attempted to dissolve the threat by encouraging scholars to appropriate and transform certain Sufi features.[39] Ibn-Tufayl undertook to do the same, not only integrating theology and Sufism but also reconciling both with his own philosophical tradition, which he never really deserted. Al-Marrākushī describes Ibn-Tufayl as "expert in all the branches of philosophy, who had studied the work of many of the truest philosophers including Ibn-Bājja [Avempace]," exposing the foremost source of inspiration for Ibn-Tufayl. Avempace's stress on the self-taught philosopher's need for experimental exploration and a life of solitude represented the foremost source of inspiration for Ibn-Tufayl's belief in the active exploration of nature in solitude.[40]

The reception and rejection of Sufi practices in al-Andalus stood as the backdrop for Ibn-Tufayl's composition. The more immediate causes, however, remained Ibn-Tūmart's political theology of the oneness of God, Sultan Abū Ya'qūb Yūsuf's concerns about contemporary turmoil, and the need for an accessible text that would represent the political theology of the Almohads. Beyond textual sources like al-Ghazzālī, Ibn-Ḥamdīn, Ibn-Tūmart, and Avempace, Ibn-Tufayl appropriated other components that exceeded local considerations and extended to world trade and discovery and to his physical surroundings.

Tales of Lost Islands

Court philosophers not only acted as players in the heart of the political nervous system; they also sat at the crossroad of cultural exchanges. Diplomats brought gifts, travelers reported about foreign nations, and merchants brought with them luxurious goods, exposing courtiers to the curiosities of far-flung cultures. Such flow of curiosities stimulated Ibn-Tufayl to frame autodidacticism in an extraordinary world. He identifies it as the epitome of oriental philosophy, a secretive Avicennian illuminist philosophy, arguably secretly transmitted across cultures from Persia, where Avicenna lived, all the way westward to Andalusia to Ibn-Tufayl. The notion of oriental wisdom, however, originated in the farthest reaches of Southeast Asia and reached back into a long process of cross-cultural exchange.

Ibn-Tufayl sets the story of Ḥayy Ibn-Yaqẓān on a legendary utopian island located in what seems to be a no-place: "Our forefathers, blessed be their memory, tell of a certain island among the Indian islands south of the equator, an island wherein people are born without a mother or a father, and wherein a tree bears women-fruit, and this is what al-Mas'ūdī indicates is the island of Wāqwāq." Ḥayy is a product of the island's special meteorological conditions, "for that island, of all places on Earth, had an Air most temperate and perfect, which is the influence of the supreme light that arises upon it . . . For there are of them who affirm, and absolutely conclude, that Ḥayy Ibn-Yaqẓān was of those, who in that region are born without mother or father."[41]

Ibn-Tufayl presents the possibility not only of spontaneous generation but also of life beyond the latitudinal boundaries of known civilization. As he suggests, what seems to be his no-place island, however, had already been mapped by the medieval geographer and traveler, al-Mas'ūdī (d. 956), in his canonical geography, *Meadows of Gold* (*Kitāb murūj al-dhahab*). Al-Mas'ūdī mentions Wāqwāq as lying at the edge of civilization in the East, "just as the Sea of China ends . . . at the land of Sofala and Wāqwāq, a country which produces plenty of gold and other marvels. The climate there is warm and the land fertile."[42] Ibn-Tufayl picks up on this theme, describing islands south of the equator as having the perfect climate to allow for Ḥayy's spontaneous generation: "As it is demonstrated in astronomy, in these parts of the Earth, south of the equator, the Sun is vertical unto the heads of its inhabitants only twice every year, when it first enters Aries and Libra. But through the rest of the year, for six months it declines toward the south, then for six months toward the north. So that they have neither excessive heat nor excessive cold, but for that cause they enjoy an equal temper."[43]

Al-Mas'ūdī never traveled to the eastern edge of the world, nor did he visit Wāqwāq. His account is based on a thorough study of geographical and travelers' literature, from which he appropriated his description. Significantly, al-Mas'ūdī mentions neither the "tree that bears women-fruit" nor the spontaneous generation of human beings and animals. Rather, it was al-Mas'ūdī's sources, which Ibn-Tufayl might also have had at his disposal, that mentioned these more spectacular claims.

Tenth-century travel journals of captains, sailors, and merchants showed the growing medieval excitement about the exploration of the south Indian Ocean and the China Sea. Sailors and merchants sought maritime routes to China and India as well as adventure in finding lost treasure, leading them to the eastern edges of the Indian Ocean. They then reported what they encountered back to geographers, stimulating new geographical descriptions of the world.

The earliest mention of the island of Wāqwāq comes from eighth-century Chinese sources that claim the island grew trees that bore children as fruit.[44] The first Arabic mention of Wāqwāq, however, comes from *The Book of Wonders of India* (*Kitāb 'ajāib al-hind*), first written at the mid-tenth century and later supplemented.[45] Authorship is traditionally ascribed to a Persian, Buzurg ibn-Shahriyār of Ramhurmuz, himself a sea captain. Buzurg collected stories from other captains and merchants of the ports of Siraf, Basra, and Uman: tales of wonder filled with remote islands piled with gold, inhabited by people with animals' legs, where trees grew leaves in the shape of men's faces and women hung down from trees by their hair. In one account, Buzurg relates, "Muhammad, Ben-Bābishād has told me that according to reports from people who had touched on the country of Wāqwāq, there is to be found a large tree with round leaves and stalks which bear fruit, similar to the pumpkin, but larger and offering some resemblance to the human face."[46] The book gives other accounts and testimonies of sailors and merchants who reported other wonders: "In Serendib they found the footprint of Adam after his fall," while on other islands they saw "flying scorpions," "monstrous creatures called tannin," and "human beings who change their sex."

The Book of Wonders of India and the geographical accounts it relied upon were shaped at a time when Islamic civilizations established trade with the Far

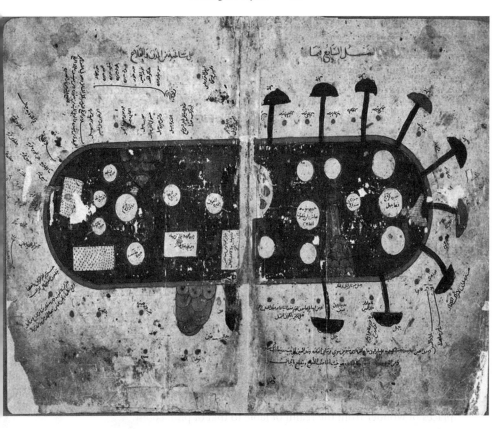

A twelfth-century manuscript illustrating the Muslims' medieval-age discovery of the Indian Ocean. The island of Wāqwāq appears in the upper left corner. From *Kitāb gharā'ib al-funūn wa-mulaḥ al-'uyūn*, known as the *Book of Curiosities*. Bodleian Library, MS Arab., cod. 90, 29b–30a. Courtesy of the Bodleian Library, Oxford University.

East and the countries around the Indian Ocean. From the eighth to the eleventh centuries, Muslim merchants expanded their commerce along all the main routes of the Indian Ocean, establishing trading posts and colonies along the coasts.[47] In this trading zone, the Muslim diasporic trading communities consisted of groups of Sufi orders that also acted as networks of trust within which Sufi merchants could smooth over any friction.[48] Sufi sheikhs held the authority to resolve conflicts, leading to a nonviolent missionary expansion of Islam. On their return, Sufi merchants brought back mystical stories about the wonders of nature and descriptions of south Asian religious practices of Hinduism and Buddhism. In bringing home accounts of remote cultures and civilizations, they infused Islam with oriental traditions of meditation, repetition of words, and ecstatic practices.

Trade and cross-cultural exchange brought Sufis to the edges of the known world, directly affecting their spatial consciousness. Arab geography flourished. Whereas early Arab geographers looked to Ptolemy's *Geographia*, they did not accept its arguments that the Indian Ocean was landlocked but believed it to be open and connected to other seas, especially in those traditionally uncharted areas south of the equator. Muhammad ibn Mūsā Al-Khwārizmī (d. 850), whose geographical work, *The Book on the Shape of Earth (Kitāb ṣūrat al-arḍ)*, gave a revised and completed version of Ptolemy's *Geographia* and included more accurate maps, was one of the first to argue that the Indian Ocean was open to other oceans. By Ibn-Tufayl's time, the question of the landlocked Indian Ocean had been all but resolved by Abū Rayḥān al-Bīrūnī (d. 1048), who first proposed that the southern sea connects with northern waters through a gap in southern Africa.

But Ibn-Tufayl had other, older sources at his disposal. His literary and philosophical use of the exceptional descriptions of the island of Wāqwāq acts as a reminder of geography's long tradition within the genre of fiction.[49] The Greek tradition of fabulous islands outside the Pillars of Heracles gradually developed into the myth of Atlantis in Plato's *Timaeus*. The developing disciplines of geography and meteorology that divided the world into climatic zones raised these places of wonder from the realm of fable to that of philosophy.

In the classic system, Earth was divided into seven climatic zones. This geographical tradition predates the idea of a spherical earth and goes back to Pythagoras and ancient Persia. Their delineation from south to north drew on the fact that these *klimata* were measured by the angle between the axis of the celestial sphere and the horizon. Later, Ptolemy derived his geography from ancient tradition and divided Earth into seven climates. Aristotle, on the other hand, in the *Meteorology*, divided Earth into five zones, with two frigid climes (arctic and antarctic) around the poles, an uninhabitable torrid clime near the equator, and two temperate climes between them. Ptolemy's system was adopted by Muslim geographers like al-Bīrūnī and Abū 'Abdallāh al-Idrisī, whereas Aristotle's remained more common in Europe.

The division of the world into climates—especially the assertion of inhabitable zones—has important cultural implications. Belief in the inhabitability of the southern temperate zone generated both mystic and Pythagorean notions of a parallel, antipodean world where cities, cultures, and religions all flourished in direct geographical opposition to that of the North. They became a cultural symbol for an alternative world, an otherworld, used to subversively criticize political states. In ethnologic satire, stories about utopian islands—such as Thule, Taprobane, and Wāqwāq—arose, in which ideals about the roles of culture and nature could be played out.

During the period of Rome's greatest expansion, interest focused on the mysterious Taprobane (Sri Lanka), which became a potent symbol as the antipode of Greece and Rome, a utopian space that could redeem the flaws of contemporary civilization. There, idealized peoples mocked the flaws of the putatively more advanced Mediterranean culture.[50] Ptolemy depicted Taprobane as an Indian Ocean island of nearly continental size and located it athwart the equator in the final regional map of his *Geographia*. Hipparchus saw Taprobane as a new hemisphere, while Pliny observed that only in Alexander's time did people come to understand that it was an island; until then people believed it to be another world. He described Taprobane simply as a land "banished by nature outside the world and hence free of the vices that plague other countries," where, in opposition to Rome, "nobody kept a slave, everybody got up at sunrise and nobody took a siesta in the middle of the day; the price of grain was never inflated; there were no law courts and no litigation."[51] Similarly, Wāqwāq is portrayed as the antipode of the al-Andalusian world: a perfect parallel space where Ibn-Tufayl could set his alternative program for an independent study of philosophy without formal education.

From the Eastern Indian Ocean to Western al-Andalus

Although far from the Indian Ocean, al-Andalus, which lies on the southern coast of Spain, constituted part of the network of trade that tied the Eastern world to the Mediterranean. We find indication of such a link in the Geniza, the medieval depository of the Jewish community of Cairo, which held dozens of documents describing two trading families who moved between the Indian Ocean and al-Maghrib in the mid-twelfth century.[52] One man, Halfon ben Nethanel of Cairo, traveled widely, returning in the spring of 1134 from India to Aden; one year later found him in Cairo, Spain, and al-Maghrib. Other North African merchants traveled just as widely, some making the long journey from their homeland to India more than once in their lifetimes.[53]

Various Andalusian sources attest to the area's exposure to information from the East, particularly characterizing the feminine attributes of Wāqwāq. One of the more fanciful appears in *The Book of Wonders of India*, which tells a story of an al-Andalusian sailor, originally from Cádiz (in southern Iberia), who sneaks onto a ship sailing for the eastern parts of the Indian Ocean. He finds himself on a remote island called Wāqwāq, where the ratio of men to women ran to "a thousand women or more to every one man. The women carried them [sailors] away toward the mountains and forced them to become the instruments of their pleasure. One after the other the sailors dropped off and died of sheer exhaustion. Only one man survived, and he was the Andalusian, of whom a single woman had taken possession."[54]

Geographical sources in Ibn-Tufayl's close surroundings represented yet another source of information about the wonders of the East. For example, the twelfth-century Sicilian geographer, Muhammad al-Idrisī, who was born in Almoravid Ceuta in southern Iberia, just a few miles from Ibn-Tufayl's hometown of Guadix, mentioned Wāqwāq by name but not the story of the "women tree." Only one al-Andalusian geographical source during Ibn-Tufayl's lifetime referred to that: *The Book of Geography* (*Kitāb al-jughrāfiyā*), written in the twelfth century, places the island of Wāqwāq in the China Sea, where trees "bear fruit that end in the feet of young girls . . . They are suspended by the hair and their form and stature are most beautiful and admirable . . . and when they fall on the ground they utter two cries 'Wāq, Wāq.' The island is at the end of the inhabited world."[55]

Ibn-Tufayl took this description of women in a radically new direction, lifting the idea of naturally grown human beings out of the realm of wondrous literature and claiming such wonder for philosophy by transforming the tree that bears women as fruits into the spontaneous self-generation seen in *Hayy Ibn-Yaqzān*. Whereas the various descriptions of Wāqwāq introduce nature as a feminine organ, Ibn-Tufayl portrayed Hayy as an active male force who conquers nature.[56]

Hayy Ibn-Yaqzān, then, came out of the medieval fascination with the Indian Ocean. The existing accounts of Wāqwāq allowed Ibn-Tufayl to present the possibility of spontaneous generation. On Wāqwāq, which lies in vicinity of "Serendib, where one could find the footprint of Adam after his fall," he presents Hayy as the first man, the quintessential *tabula rasa* who, like the biblical Adam, explored and conquered nature without recourse to philosophical tradition, genealogies, literacy, or genetic history. The wondrous descriptions of Wāqwāq supplied Ibn-Tufayl with an experimental space where he could stage a story in the ancient tradition of ethnologic satire to subvert textual and religious authorities. By setting *Hayy Ibn-Yaqzān* at the ends of the known world, in a space already colonized by fabulous literary tales, Ibn-Tufayl transformed the story into a thinking experiment.

A fourteenth-century illustration from *The Book of Wonders* describing the most notable curiosity found on the island of Wāqwāq: women hanging from a tree that came to be known as the Tree of Wāqwāq. *Kitāb al-bulhān*, Bodleian Library, Or. 133, fol. 41v. Courtesy of the Bodleian Library, Oxford University.

Ibn-Tufayl inserted other local features into the narrative of *Ḥayy Ibn-Yaqẓān*, features that went beyond philosophy and theology to incorporate the physical space in which he lived. At the climax of the story, Ḥayy "restrained himself, in the lowest part of a cave," to meditate and "then he deeply got plunged in the vision of *that first being, the true necessary existent*" from which all species arose. For a communion with the "first being" Ibn-Tufayl chose the cave rather than the beach, under a tree, or under the stars—the other locations of Ḥayy's practical contemplation. A connection to Plato's cave or other Islamic traditional stories can easily be seen, but Ibn-Tufayl's choice stems from a structural association to a much closer source: Guadix, his hometown in southern Iberia.

Totemic Taxonomy

Ibn-Tufayl's choice of the nursing mother-gazelle was not the mere fruit of creative writing. He had the philosophical gazelle, al-Ghazzālī, to which he related as a starting point of his autodidactic philosophy. He also, however, had cultural and historical symbols in his immediate surrounding that served as the raw material for his tropes. Such symbols and their interpretations come to light not through traditional textual evidence but by means of archaeological remnants and anthropological rituals. The Iberian Peninsula has been deemed the place where European history began, its many prehistoric archeological sites contributing to that widely held belief. Archeological evidence from prehistoric caves indicates that southern Spain acted as the bridge over which human beings first moved from Africa to Europe. The southern coast of the Iberian Peninsula stood at the center of these prehistoric caves.[57] Archeologists had found paintings of animals, including the female gazelle, in some celebrated prehistoric settlements and caves near Guadix—Los Millares, Nerja, Casares, Horá, and Chimenesas. Ethnographic studies of contemporary hunter-gatherer societies stress that the paintings were made by shamans who retreated into the darkness of the caves, entered into a trance state, and then painted images of their visions.[58] Not only were prehistoric symbols deployed around Ibn-Tufayl's habitat. Guadix itself is now, and was then, actually a village of cave homes.

Cave paintings of animals' forms were not simplistic expressions of figurative art. They functioned rather as natural and mythological symbols, or totems, whose energy might speak to an individual in a way that was relevant to his personality. The gazelle, for instance, symbolizes awareness in shamanic religions, and in using this trope Ibn-Tufayl further insinuated the principle ingredient

Prehistoric totemistic art. Paintings of totems like gazelles, oxen, and other mammals were widespread in caves along the Iberian Peninsula and especially near Gaudix, the hometown of Ibn-Tufayl. Replica of a painting from Altamira Cave in Spain. Courtesy of National Museum and Research Centre of Altamira.

for the autodidact—awareness—as it was disclosed in the literal meaning of the name Ḥayy Ibn-Yaqẓān—Alive Son of the Vigilant.

Through such totems prehistoric man rearranged and classified the natural world. The anthropologist Claude Lévi-Strauss argues in *The Savage Mind* (*La pensée sauvage*) that totems were chosen arbitrarily for the sole purpose of creating a comprehensive and coherent classificatory system of the physical world. Through totems, prehistoric man ventured for the first time to understand the order of nature—natural objects were named, their qualities were classified, rearranged, and attached to single social functioning. Lévi-Strauss distinguishes modern scientific taxonomies that claim to be detached from social relations, and thus objective and universal, from totemistic taxonomies that are based on "direct sensible intuition"—the organizing principle and the principle practice of the "science of the concrete."[59] For Lévi-Strauss, totemistic taxonomy laid the infrastructure for our civilization.

Ḥayy Ibn-Yaqẓān, then, stands for more than just an ahistorical wild prodigy—generated spontaneously, living self-sufficiently, discovering nature through totemistic taxonomy, and reaching the ultimate knowledge in a dark cave, all attributes and practices correlated with the features of prehistoric man. Ibn-Tufayl not only merged mysticism with philosophy, he also fused ancient local rituals and totems with contemporary philosophical procedures, positing that the first and the quintessential autodidact was the first man, prehistoric man, or, in his baptized form, the biblical Adam.

The crossings of al-Andalusian currents in the twelfth century stimulated discussions about autodidacticism. Sufi arguments for the individual and his direct experiencing of God conflicted with the al-Andalusian philosophical tradition that relied heavily on logic and mathematics. Almohadi political theology advocated for religious accessibility and thus revolutionized the old jurisprudence tradition, which relied on texts that allowed only judges to determine and interpret the premises of theology. Instead of words, Almohadi officials and philosophers stressed the importance of individual action and thus promoted the notion of personal responsibility as a political-theological concept that could facilitate the centralization of the Almohadi kingdom. The Andalusia of the twelfth century was no longer a removed western province of the Islamic world but was now connected to streams in the circulation of philosophies that came from the Arab East and also to books of wondrous stories written by sailors who sailed to the southeastern Indian Ocean. The stories told of utopian island societies that acted as medieval laboratories where spontaneous generation replaced the sociobiological notion of genealogy and where firsthand experience replaced traditional and authoritative transmission of knowledge.

The crossings of such currents yielded various cultural symbols that Ibn-Tufayl had at his disposal when writing his own accessible philosophical text. In *Ḥayy Ibn-Yaqẓān* he reconciled such contradicting currents by placing the story on a Far Eastern utopian island, a no-place island named Wāqwāq, just below the horizons of climates and civilizations, where the laws of nature and society no longer applied. There Ḥayy became a first man, a quintessential autodidact, who builds up his philosophy through the practical exploration of nature. He also represents the ultimate pious Almohadi, who turns ethics, personal experience, and responsibility into the foundation of his religious life. In his nonteleological, inductive search for the necessary being, he subjects the Sufi science of practice to philosophical procedures and explores nature through inductive reasoning. As the superior creature in nature, the lively, resourceful Ḥayy conquers and takes control of the passive nature surrounding him.

Although Ibn-Tufayl's *Ḥayy Ibn-Yaqẓān* was a short tale, it became a source of inspiration for other stories about wild autodidacts. In the thirteenth century, Ibn al-Nafīs wrote a theological story, *The Complete Treatise on the Prophet's Biography* (*Al-Risāla al-Kāmilīyah fi al-sīrah al-nabawīyyah*), which also told of a feral child living on a desert island, but the plot later expanded beyond this setting and evolved into a kind of science fiction story. In the early twentieth century, Ibn al-Nafīs's story was translated into English as *Theologus Autodidactus*.[60] But the story did not appeal only to Muslims. Its presentation of a first man who preceded revelation led the story of *Ḥayy Ibn-Yaqẓān* to be perceived as dealing with universal revelation, and so it attracted the attention of non-Muslim scholars, especially Spanish Jews, who had access to the circulating Arabic manuscripts.

Ḥayy Ibn-Yaqẓān attracted so much intellectual attention that, in 1880s-era Cairo, it became one of the first works to be printed in the Arab world. Farah Antūn, a Lebanese intellectual, the editor of the Arabic periodical *al-Jami'a* and author of several books, including a famous work on Averroes, published one of the first Arabic editions (1909). Antūn rose to fame as one of the pioneers of modern secular thought in the Middle East. As a Christian, and heavily influenced by "the French orientalist Ernest Renan, Antūn addressed the question of religion and science in Islam and the Middle East and thought about Islam's capacity to tolerate and appropriate Western science. He promoted *Ḥayy Ibn-Yaqẓān* as a means of showing Muslims that the origins of Western science actually lay in medieval Islamic philosophy.

More recently, the story has once again been transformed to serve religious and educational purposes. A cartoon version of the story, produced by an American Islamic organization, has been used for the education of boys according to ascetic Islamic values. Ḥayy, ultimately, has become a cartoon hero, a religious Islamic wild boy, in contemporary Arabic culture.

CHAPTER TWO

Climbing the Ladder of Philosophy
Barcelona, 1348

As ARABIC PHILOSOPHICAL TEXTS moved from Spain, through Catalonia and across the Pyrenees Mountains into France and Provence, numerous translations into Latin and Hebrew—including of *Ḥayy Ibn-Yaqẓān*—began to appear, sparking local controversies between Christian philosophers and theologians at the University of Paris and among Jews in the communities of Provence-Catalonia. More than a century later, Moses Narbonni, a Jewish physician and philosopher, restlessly wandered through Provence Catalonia. When, in 1348, he arrived in Barcelona, he set out to write a commentary on the anonymous Hebrew translation of *Ḥayy Ibn-Yaqẓān*, titling it *Yehiel Ben-'Uriel*.[1] This commentary aroused intellectual interests in autodidacticism and became the vehicle through which later generations of Hebrew and Latin readers first came to know *Ḥayy Ibn-Yaqẓān*.[2] Cultural circumstances, however, stimulated Narbonni to engage with the text. For decades, an ongoing controversy regarding whether or not adolescents should study philosophy swept him and his philosophically inclined community in Perpignan into war with leading legal scholars in Barcelona—with excommunication as the ultimate threat. Under such social pressures Narbonni wrote the commentary, celebrating Ḥayy Ibn-Yaqẓān, the boy who became a philosopher; introduced autodidacticism as the best philosophical program; and proposed that adolescents represented the finest candidates for its practice.

Having arrived at Barcelona, Narbonni found his way to the Jewish quarter, *la juderia*, or in Catalan language, *call*, entering its narrow streets through the gate that was located near the cathedral, between La Plaza de Sant Jaume and Sant Honorat. He introduced himself as a physician and Maimonidian philosopher who was born, educated, and trained in Perpignan, a little town beyond the Pyrenees, on the border between Provence and Catalonia. Narbonni's self-introduction, however, stirred strong estrangement. Although Perpignan was only fifty miles northeastward, it represented a radically different cultural worldview. Counterpoising differences between the two local cultures stemmed from the various political contexts that shaped particular structures of the communities.

The old Catalonian communities carried on their old traditions. In internalizing the continuous conflict between Christians and Moors they sought to secure the character of their community by tightening up the cultural boundaries and confining intellectual activity to the *juderias*. Communities of Provence, on the other hand, benefited from the growing need for commercial centers in late medieval times. Local rulers offered incentives for Jews to move to their towns. James I of Aragon, who ruled Perpignan during the years 1213 to 1276, granted many privileges to the Jews, and so the population drew new members from surrounding towns. Young and fresh Jewish intellectual centers came into being in late medieval Provence. As the regional scale of trade increased, Jews moved to towns like Lunel, Montpellier, Narbonne, and Perpignan, connecting Provence to networks of trade across the Mediterranean. Whereas the Jewish communities in Catalonia were deprived of rights and marginalized, Jews in communities of Provence enjoyed the same privileges as their Christian fellow citizens, but they also bore the same fiscal responsibilities.[3] As the Catalonian communities closed the cultural gates and retreated to maintain their hierarchical semiautonomous community, the Provençal communities openly exchanged with their surroundings, enabling grassroots cultural activity to develop.[4]

Each community was regulated by a board of elders. The royal or seigniorial authorities recognized the *episcopus Judaeorum* (bishop of the Jews) and a board of elders, the *seniors,* as representatives of the community who were elected annually. The *seniores* (in Hebrew, *zekenim* [elders or old wise men]), were elected to the board by virtue of their life experience. Lacking life experience, young members of the community were expected to receive the store of their elders' knowledge and to follow their regulations on everything from scholarly agendas to social functioning. The *seniores* regulated social life from top to bottom, directing local pedagogy along traditional religious lines. Most children were educated in *heder* (Hebrew school), where small groups of adolescents learned basic Hebrew and religious studies. In Provence the *seniores* were not as anxious for their authority, encouraging extracurricular education. Wealthy families further invested in education by hiring private teachers of a wide range of subjects, including Greek and Arabic philosophy. Success in study brought with it exposure to the surrounding Christian society, appointments to political positions in courts, and in some cases even conversions to Christianity.

The structures of the communities further correlated with their approaches to teaching philosophy. Whereas older communities with strong local tradition, such as Barcelona, Valencia, and Cervera, maintained their hierarchical structure through strict education by elders and viewed the study of philosophy and humanist subjects as a step toward assimilation, the fresh Provençal communities of Montpellier, Narbonne, and Perpignan encouraged adolescents to study philosophy without recourse to elders' guidance.

In the lack of roots and local tradition, communities such as Perpignan struggled to moderate the tension between tradition and experimental cultural trends, allowing philosophical education to more easily take hold. Narbonni's biography, gleaned from the manuscripts he left behind, tells of a boy, born toward the end of the thirteenth century, who was philosophically educated in openminded Perpignan and who ended his life lonely in "exile" in the communities of Catalonia, around 1362.[5] All but one of his writings were commentaries, a form uniquely suited to the pedagogical task of providing a didactic introduction to the major philosophical issues to readers for whom primary philosophical sources remained hard to grasp.[6] Fluent in Latin, Castilian, and Provençal French, he had only a basic knowledge of Arabic, preferring to use Hebrew translations of Arabic sources for his commentaries. Probably educated by private teachers in Perpignan, he began studying Moses Maimonides' *Guide for the Perplexed* at the age of thirteen with his father.[7] He signed his manuscripts with his Hebrew name, Moshe Narbonni, but state documents in Barcelona refer to him by his Latin name, Maestro Vidal Bellshom de Narbonne.[8] Although his name derived from the city of Narbonne, an important center of Jewish life in Provence between the eleventh and the fourteenth centuries,[9] he explicitly referred to Perpignan as his hometown. The community books in Perpignan indicate that a few families named Narbonni emigrated from Narbonne sometime in the second half of the thirteenth century.

Narbonni lived in Perpignan, where he completed the writing of a few commentaries, until 1344. Years before he started writing the commentary, Narbonni had lined up *Ḥayy Ibn-Yaqẓān* as a project he wanted to pursue. In his commentary on Averroes' epistle, *The Possibility of Union* (Efsharut haDevekut), completed in the 1340s in Perpignan, Narbonni states his plan to compose a commentary on *Ḥayy Ibn-Yaqẓān* "so to describe the regime of solitude and the way to achieve a communion with God."[10] From Perpignan he moved to Cervera, which he later refers to as "the place of my exile" or "the place where I was exiled to."[11] The move from Perpignan to Catalonian communities, however, symbolizes not simply a change of location but also a shift from a bottom-up social structure and

Frontispiece of a decorated manuscript of Maimonides' *The Guide for the Perplexed*, produced in Barcelona in 1348. Kongelige Bibliotek, Copenhagen, cod. Heb. 38.

open-minded culture to a more hierarchal structure and conservative cultural view.

When Narbonni arrived at the unwelcoming Barcelona, he took part in stimulating local scholars to produce a magnificently illustrated copy of Maimonides' *Guide for the Perplexed* and began to write his commentary on *Ḥayy Ibn-Yaqẓān;* but the conservative intellectual tradition as well as natural disasters limited the duration of his stay in the city. In 1348 a plague hit the city full bore, and thousands quickly lost their lives. Blame fell on the Jews, and soon rioting mobs stormed the Jewish quarter and massacred hundreds. The tragic events cut short his work, and he had to escape the city. He returned to Cervera, where he completed his commentary, entitling it *Yehiel Ben-'Uriel,* and attached to it a commentary on Avempace's *The Regime of Solitude.*[12] The introductory paragraph to *Yehiel Ben-'Uriel* says that the community of Perpignan stirred him to write the commentary and to make its arguments accessible to everybody so as "to wake up those who are asleep."[13]

Of the themes presented in *Ḥayy Ibn-Yaqẓān,* Narbonni choose to accentuate the defiance of traditional authorities, the advantages of the solitary life, and the urge to self-explore nature and God, showing that anybody, even a lonely boy, can climb the ladder of philosophy—from physics to astronomy and metaphysics, and only at the end to theology. The blend of such philosophical themes with the Perpignans' request for a commentary suggests that *Yehiel Ben-'Uriel* aimed to provide an answer to a collective problem. The problem was not new: some decades earlier, Asher Ben-Yehiel, a leading scholar of Jewish law and a strong opponent of philosophy, passed through Perpignan on his way to Barcelona from the German-France border. Observing the passions stirred up by the teaching of philosophy in Perpignan and the public discussion of philosophy in the town's synagogues, he sarcastically stated that "only a few Jews were left there."[14] Ben-Yehiel's comment not only encapsulated the cultural differences between the conservative Catalonian and the open-minded Provençal communities. As philosophy increasingly trickled down to cultural practices, Ben-Yehiel's words also fueled a forthcoming political clash over philosophy and education.

"To Wake Up Those Who Are Asleep"

Philosophical studies grew stronger in the communities of Catalonia and Provence as Jews escaped the persecution of the Almohads in the late twelve century or the Muslim-Christian war zones. They found their way north to the adjacent parts of the western Mediterranean, bringing with them the linguistic skills and sincere

Title page, Narbonni's commentary, *Sefer Abu-Baker: Hayy Ben-Yoktan,* in the center a gazelle lying under the tree. Bayerische Staatsbibliothek, Munich, cod. Hebr. 59.

intellectual commitment necessary for an intense engagement with Arabic philosophical writings.[15] Scholars had begun translating the Arabic text into Hebrew, expanding the scholarly circles and making philosophy accessible to locals who could not read the Arabic sources. When the writings of Maimonides arrived in Provence, they sparked various intellectual controversies that lasted more than a hundred years, from the early thirteenth through the early fourteen centuries— the period of Narbonni's birth and upbringing.

The controversies over the reception of *The Guide for the Perplexed* revolved around metaphysical and political questions, particularly concerning the correlation of the laws of nature with the laws of society. Just as the order of nature reflects not only the omniscience of God but also his active and constant rule, so too with a political ruler, who should be not only a passive lawgiver but also an active political leader. Since Jews lacked political rule, the focus turned to the relation of a scholar with his surrounding community.

For Maimonides, the biblical story of Jacob's dream of the angels' climbing up and down on a ladder that reaches the heavens exemplified the role of the philosopher, who must combine the contemplative with the active life. He must climb the ladder of philosophy through the exploration of physics, mathematics, logic, and metaphysics, reaching its uppermost divine level and then returning to society for the purpose of securing the well-being of the community. This division also represented the everyday life of the philosopher, who should work and make a living, dealing with philosophy only during his free time. In a letter he wrote from Egypt in 1199 to Shmuel Ibn-Tibbon, his translator in Provence, Maimonides personified his stance and described how he combined the two lives. In the daytime he practiced medicine and worked on social issues concerning his community; only after "laying down from exhaustion" at night was he available to contemplate and study. Sometimes, when his active social life consumed his time, he "would run away from human company and unnoticeably dedicate his time to writing."[16]

Maimonides used a verse from the Song of Songs (5:2), "Ego dormio et cor meum vigilat" (I sleep, but my heart awakes) to encapsulate the ideal relations between the contemplative and the active life. While the practical life represents a state of intellectual sleep for the ordinary philosopher, the contemplative state of cognition in an ideal prophet-leader is similar to that of biblical Moses, the prophet-philosopher-leader.[17] Even when the philosopher sleeps (that is, is active in social and political spheres) his heart remains vigilant; his mind still points to spiritual and philosophical endeavors.

Some of his followers in Provence-Catalonia promoted his philosophy but did not understand the verse from the Song of Songs as Maimonides intended. For example, Shmuel Ibn-Tibbon, who set off the reception and dissemination of *The Guide for the Perplexed,* took a different stance. In his commentary on Ecclesiastes, he writes that Moses' social and political activities could have borne more fruit had he been obeyed more often. Ibn-Tibbon went so far as to propose that had Moses dedicated his time wholly to contemplation, he would have achieved even more intellectually. Seeing a parallel, Ibn-Tibbon claimed the same for Maimonides, who had worked hard to lead his community but whose books were eventually burned in France. And if the foremost prophet, the biblical Moses, and the foremost Jewish philosopher, Moses Maimonides, could not reform their societies, others had even less hope of attaining such ends.

Therefore, Ibn-Tibbon held, a philosopher should focus on the "awakened heart," the philosophical life, and give up social and political vocations. Ibn-Tibbon concludes by positing a difference between "one who is fast asleep and his heart awakes, and one who is completely awake," asserting that the perfect man who tries to do both cannot exist outside a utopian philosophical community. Instead, he presents the possibility of human perfection in the model of Noah and Enoch, quintessential solitaries who were "wakeful contemplators," who lived in exile within their societies and so did not fall when their surroundings did. The rejection of social constraints and intellectual authority also stressed that the perfect intellectual courts autodidactic education and solitude.[18] The diversified stances regarding philosophy that ranged from Maimonides' "social philosopher" to Ibn-Tibbon's "isolated philosopher" represented two approaches that stood in sharp contradiction to the conservative intellectual culture of Barcelona.

Born into Controversy

Narbonni was born into the stimulating intellectual culture of the late thirteenth and early fourteenth centuries, when philosophy and Kabbalah went public and stimulated cultural unrest. What had been considered esoteric knowledge, available exclusively to a small circle of scholars, became more and more accessible to more and more people as Kabbalist works were brought to light by scholars eager to expose a latent current of ancient wisdom. Similarly, scholars in Provence-Catalonia, looking to Maimonides for inspiration, undertook the translation and cultivation of Greek philosophical works from Arabic sources to synthesize them with Jewish theology, law, and hermeneutics. Their natural audience lay not with the tradition-bound elders but with the younger generation, who were open to this new way of thinking.

The result of this synthesis was a popularization of philosophy during the late thirteenth and early fourteen centuries. In rabbinical *responsa,* or written legal literature, of the time, central rabbis incorporated philosophical texts and concepts into their replies to their followers. Commentaries on the Bible, especially on the book of Genesis and the story of the creation, also reflected new ways of thinking, by trying to apply Aristotelian physics to the story.[19] Local readers began to view the Bible through the lens of philosophical questions about cosmogony and metaphysics. At the same time, an oral philosophical tradition developed when scholars started giving public speeches on philosophy, which were also recorded so that others could have access to them. The expansion of the appeal of philosophy, from the traditional locus of authority—the elders—to the general populace—the young and the nonscholarly—inevitably headed for a clash. In the early fourteenth century, it erupted.

Public speeches connected intellectual agendas with cultural practices. During the thirteenth and fourteenth centuries, noted rabbinic scholars delivered speeches at communal gatherings. Community events—such as Sabbath and holiday services, commemorative meals, celebrations of weddings, and bar mitzvahs—all represented opportunities to address the collective consciousness. At the same time, these cultural events established the scholarly and political authority of the local leadership. Wealthy leaders sponsored the events while scholars provided the entertainment, in the form of intellectual speeches that both introduced the fruits of their intellectual labor and justified the cost of the public investment in their stipends. The scholar-speaker was expected to make philosophy accessible to the public. By breaking down esoteric fences and using allegories and metaphors, he could convey abstract notions and ideas. While these events represented opportunities to assert authority, however, they also carried the potential to challenge that very authority. Public speeches, therefore, were the quintessential vehicle for the popularization of philosophy—given orally, in colloquial language, using cultural symbols and metaphors familiar to the audience.

Sometime in 1303, a speaker at a Jewish wedding allegorized the biblical figures of Abraham and Sarah as philosophical concepts of matter and form, igniting a controversy that climaxed around 1305 and flamed for decades. In response, Aba Mari of Lunel, a scholar from Montpellier who combined interests in philosophy and religion, wrote to the leading scholar of Jewish law, Rabbi Shlomo Ben-Aderet (1235–1310) of Barcelona, who already had strong reservations about Maimonidian philosophy. In the first of many letters (later collected in book form and titled *Sacrifice of Zeal* [*Minhat kanaut*]), Mari appealed to Ben-Aderet's conservatism

A public speech in a synagogue, from a Hebrew manuscript from Catalonia illustrates the public reading of the Haggadah, made around 1350, a widespread practice in northern Spain and Provence. In casting philosophical arguments in biblical stories, Provençal scholars sparked the controversy whether everyone, especially adolescents, should be exposed to philosophy. British Library, Or. 2884, fol. 17v.

and urged him to stop the "irresponsible spread" of philosophy, expressing his chief concern over the shaking up of the traditional authority of the elders. "This conceited young generation dares to judge their own judges and elders," he complained, "to compose some books on logic and natural philosophy and to turn the books of Averroes and Aristotle to their pillars of belief." The study of philosophy became a threat to tradition, since those who philosophize "break down the walls and fences and pass on [their philosophical ideas] to children." He exhorted Ben-Aderet to "put your sword on your waist, take a stick and hit them on their heads."[20] Unlike Ben-Aderet, though, Mari was not against philosophy itself; rather, he wanted to return the study of philosophy to the confines of a close circle of philosophers and to keep it away from "irresponsible" adolescents.

Because of the nature of education at the time—given by private teachers in private houses—the elders had little recourse to stop the spread of philosophy by the independent scholars who gave these lessons. Such education stressed that the fundamental principles of religion can only be understood through an acquaintance with the philosophy of nature and logic. This, of course, represented a clear subversion of traditional law studies and the authority of the elders of the community. The only way left for the elders to guard their authority over curriculum, and in the process "build walls" to fortify tradition, protect the community from assimilation, and perpetuate scholarly and social authority, was through excommunication, which could be issued only by a senior legal scholar supported by other rabbis.

Shlomo Ben-Aderet was one such scholar. In 1303 he issued a famous excommunication against those who studied and taught philosophy. In his ruling, he writes that "using the force of excommunication we determine that nobody from our community shall study the books of the Greeks on physics and metaphysics, either in their language or copied into other languages."[21] The legal ruling threatened to excommunicate anyone under the age of twenty-five who studied philosophy and those who taught them the banned material. It explicitly limited the study of philosophy to a small circle of men and forbade teaching it to adolescents without strict pedagogical guidance. Furthermore, works like *Ḥayy Ibn-Yaqẓān* were specifically targeted. Later in the same document, Ben-Aderet writes that "a boy who is born and raised in the lap of natural philosophy would be taken overwhelmingly by Aristotle's demonstrations and then turned into a complete heretic."[22]

The metaphors deployed in connection with the ruling of excommunication expressed the grave nature of the threat to the community's sense of itself. The messengers from Barcelona who reached the communities of Perpignan, Lunel, and Montpellier carried with them the ruling, in which the study of philosophy was compared with "marrying a gentile woman" or "publicly kneading the breasts of a gentile woman," associating philosophy with dangerous temptations of the gentile—that is, alien—woman.

Ben-Aderet, Mari, and others justified their opposition to the study of philosophy by asserting the corruptive affect of exposing impressionable adolescents to heretical opinions and methodologies. They also argued that the study of Jewish law, a law that regulated everyday life and maintained the structure of the community, suffered neglect when philosophy remained on the curriculum. Ben-Aderet argues that philosophy "breaks down the wall and the fence that protect and preserve the community," especially for those whose "opinions are imperfect and naive, like little boys and women . . . who might by philosophical demonstration think that the Torah is not wisdom, but a complete chaos." Ben-Aderet concludes that we have "to rebuild the fence and the wall by enforcing the study of Jewish law and not of cosmogony and metaphysics," about which, he emphasizes, adolescents speak in zeal "in the markets and on the streets." This argument was echoed in Perpignan. Bonfosh Vidal, a local scholar of Jewish law, wrote back to Ben-Aderet and emphasized the need to keep adolescents from studying magic and philosophy, urging him to "build fences, walls, and regulations for the adolescents of Israel, to avert them from the study of logic."[23]

For conservative scholars like Ben-Aderet and Vidal, the consequences of teaching philosophy to adolescents were devastating. "As we witness the fall of our generation," Ben-Aderet stresses the need to protect adolescents from natural philosophy by "completely excommunicating . . . those who will teach or be taught natural philosophy before the proper age of twenty-five and before filling his stomach with the rules of the Torah."[24] But the excommunication ruling did not pass without resistance.

Perpignan Fights Back

The excommunication ruling generated a counterruling of excommunication (against whoever followed the Barcelonan ruling) led by the scholars of Perpignan. In a reply letter to Ben-Aderet, Perpignan-based opponents observed that Maimonides incorporated philosophy into Jewish law, hermeneutics, and theology. They noted the practical benefits of applied science—astronomy facilitates the timekeeping necessary to the Jewish calendar; geometry and engineering allow construction and fortifications; and meteorology helps create universalized measurements that standardize trade and travel and help keep the social order. They ended their letter with a strong methodological argument that boys' pliable minds are more ready to receive philosophical ideas. "You better put your sword back at your waist," they warn Ben-Aderet and conclude, "This is not our way to avert philosophy from babies and boys," since without philosophical education no mature and perfect Jewish scholars could emerge. "Without young goats," they end, "there are no he-goats."[25]

Menachem ha-Meiri (1249–1315), a leading scholar of Perpignan, opposed the efforts to marginalize Jewish law, as radical Maimonidians tended to do, or to exclude philosophy from the curriculum altogether, as legal scholars like Ben-Aderet advocated. Ha-Meiri occupied a central position, one that accepted both philosophy and Jewish law as fundamental for the education of young scholars.[26] Disallowing philosophical education for adolescents, he felt, might present the loss of a rare opportunity to use philosophy in adulthood. "Not everybody," ha-Meiri wrote to Aba Mari, "is entitled to study the secrets of wisdom, only some";[27] but everyone who is given the opportunity to study philosophy should pursue it whenever possible, since the future struggle to make a living and the lack of teachers eventually avert men from studying philosophy. The experience of the community of Perpignan, then, was not a clash between philosophy and religion; rather, they addressed considerations regarding when and who should study philosophy and to what degree philosophy should go public. But the controversy did not lie in the distant past and had not died down; it affected Narbonni's immediate surroundings and situation, as well.

Even within culturally open-minded Perpignan, however, a group of people followed the ruling. In the social and cultural turmoil stimulated by the pedagogical controversy, one side had informants that reported to Barcelona about those who taught philosophy to adolescents, and the other side fought back by issuing a counterruling. This went on in Perpignan even after the Jews were expelled from France in 1306, when Aba Mari, the major figure who originally stirred this pedagogical opposition, left Montpellier and immigrated to Perpignan. A group of his supporters received him on his arrival. Their leader, Moshe Bar-Shmuel, briefed him that "our opponents [those who promoted the study of philosophy to adolescents] put a lot of effort into convincing the King of Majorca not to admit you to Perpignan."[28]

The two camps engaged in a prolonged and nasty struggle, with the Narbonni family aligning with those who continued teaching philosophy to adolescents. Narbonni himself testified to his childhood experience with philosophical study, but more than that, extensive correspondence shows his family caught up in the midst of the struggle. In 1305 Ben-Aderet wrote to Don Crescas Vidal of Perpignan that he heard that men and boys in his town "released their bridle" and taught philosophy before the Torah. "There are men [in Perpignan]," Ben-Aderet

Ḥayy contemplating the dead body of his mother-gazelle, a moment of puzzlement that represents the beginning of his active exploration of the secrets of nature. Moses Narbonni, *Yehiel Ben-'Uriel (Ḥayy Ibn-Yaqẓān)*. Courtesy of the Biblioteca Comunale Teresiana, Mantua, Heb. 12, p. 21.

continued, "that their books should be burnt; either books of magic or books of nature." He urged Vidal to take action, to "expose your strong arm and break this trend." Then another letter came to Don Crescas Vidal, this time from his brother, Don Bonfash Vidal, who heard that in Perpignan, "those who are breaking the walls [of tradition] by teaching the books of the Greeks come from a few families." In his reply, Don Crescas named names and reported from Perpignan that "one of the vocal supporters in philosophy is native to the city and two others are originally from Narbonne."[29]

Notary documents from the thirteenth century indicate that among the four hundred Jewish families of Perpignan, seven households took the name Narbonne,[30] including Narbonni's grandfather, Mossé, and father, Yehoshu'a, whose enthusiastic support of philosophical studies echoed in Barcelona. As a result, the Narbonnes of Perpignan faced persecution at the hands of the traditional authorities of Jewish scholarship, likely including excommunication.[31]

Although the text of *Yehiel Ben-ʿUriel* reveals only an implicit acquaintance, Narbonni's other writings make clear reference to the controversy, particularly regarding his childhood, when it reached its climax. In his commentary on *The Guide for the Perplexed*, he points to an event that shaped his intellectual approach: a public controversy he witnessed during his childhood.

> In the time of my youth, when I was studying philosophy, I saw one who was raised on it and aged in reading and studying it, in my hometown Perpignan. Here there was a crowd of students who ridiculed and scorned one who gave a public speech and argued that some of the beliefs are true and some are false. And there was another religious scholar who thought that it is completely forbidden to refer to any parts of the belief as false, since it is whole truth. And his students agreed with him and called to excommunicate the philosopher who gave the public speech.[32]

The young Narbonni did not hesitate to enter into an argument and publicly debate the philosophical qualifications of religion, even if it meant incurring the social price of his stance. "Although I personally detest him [the philosopher], I detest false opinions more. I saw him being persecuted," he writes, "and for the love of the truth; I started arguing his arguments and confronting his opponents."[33]

No evidence testifying to Narbonni's last years in Perpignan, in the late 1330s and early 1340s, exists. He left Perpignan after it was conquered by Pedro IV, the king of Aragon, and began wandering among the towns of Spain and Catalonia, sustaining himself with the practice of medicine and by teaching, studying, and writing philosophy. During his travels he kept up correspondences with the people of Perpignan and along the way gathered other admirers, who looked up to him as a public intellectual. The introduction to part 2 of his commentary on *The Guide for the Perplexed* notes that the people of Seville and Barcelona also asked him for philosophical and theological opinions. Narbonni, then, debated publicly all his life, and not only with Jews: in his commentary to *The Guide for the Perplexed*, he describes a debate he had with one of the "most wonderful scholars of the Romans."[34] He even participated in Jewish-Christian polemics and wrote an *Epistle on Free Will* [Mamar be behira] against the deterministic view of Abner de Burgos, a Jew who converted to Christianity and promoted the mass conversion of the Jews as part of the process of cosmic Christian redemption.[35]

Narbonni's Long-Distance Dialogue with the Controversy

A few decades later, at the request of the people of Perpignan, Narbonni wrote a commentary on *Ḥayy*, in seeming defiance of the excommunication ruling issued in his father's time. However, he engaged the controversy not in an explicit reaction against the Barcelonan view but through the tropes and values of autodidacticism. Whereas the original title of Ibn-Tufayl's work was *Ḥayy Ibn-Yaqẓān* [Alive Son of the Vigilant], Narbonni called his *Yeḥiel Ben-'Uriel* [Long Live God, Son of the Vigilant God], rendering the translation a combination of names rather than translated nouns and adjectives. The introductory paragraph expresses the purpose of the commentary and implicitly ties it to the philosophical controversy over the verse of the *Song of Songs* (5:2), "I sleep, but my heart awakes." Narbonni notes, "I write a commentary on *Ḥayy Ibn-Yaqẓān* to wake up those who seek eternal life. The commentary shows how one can reach eternal life through awakened and active exploration of the Truth. Therefore, I called it *Yeḥiel Ben-'Uriel*."[36]

These carefully considered titles enter into dialogue with extant texts and, perhaps more significant, with the knowledge they convey. In this case, Narbonni says that he writes "to awaken those who sleep;" that is, to rouse those who have drifted into intellectual complacency. In doing so, he references the medieval tension between the value traditionally attributed to the secluded and contemplative life within a cloister and the value of the life of an active citizen who engages in commerce with the affairs of the world. Already, the wording of Ibn-Tufayl's and Narbonni's titles point in such direction. *Ḥayy* (alive) and *Yaqẓān* (vigilant) and *Yeḥiel* (living god) and *'Uriel* (awaking god) represent Ḥayy's active endeavors to explore and control nature through trial and error. In medieval Jewish intellectual culture the tension between *vita activa* and *vita contemplativa* not only represented the question of secluded versus social life but also stood for scholarly preferences. Jewish scholars considered the study of religion as contemplation of God and the study of philosophy as active exploration of nature and society.

Reinforcing the position of those who perceived the philosopher as a solitary man, Narbonni describes the island of Wāqwāq as "one of the Indian islands, south of the equator. This is the island in which man is created without parents." He goes on to identify Ḥayy with the first solitary philosopher, the biblical Adam: "The ancients remembered ... an island of the islands of India, south of the equator ... and this is the island, if you want, of the first man."[37] Even a century after the writings of Ibn-Tibbon, at the time of Narbonni's birth, the controversy still raged, especially in the geographic heart of Provence-Catalonia—and in Narbonni's hometown, Perpignan. Like many of his neighbors, he followed Ibn-Tibbon's radical interpretations but with a nuanced modification of Maimonides' emphasis on the importance of social life. Between these two philosophical poles, Narbonni presented the solitary wakeful philosopher who relies on his firsthand experience to make philosophy and whose contemplative faculties awaken because of, rather than in spite of, his engagement in practical activities.

Narbonni's notion of the solitary philosopher departed from the models he inherited. He explains that Ḥayy's search for the material necessary for his existence represents an ideal image of the way a solitary man should live. "You should know," he tells his readers, "that he [Ibn-Tufayl] relied on Avempace's *Regime of Solitude*," a book that encourages philosophers to live in exile and isolation within society and to create their own secret societies.³⁸ In mentioning Avempace's *Regime of Solitude*, he suggests a different understanding of the solitary man, one who lives in solitude or among solitary philosophers, surrounded by imperfect society. Narbonni highlighted the point in his commentary on the final section of *Ḥayy Ibn-Yaqẓān*, when Ḥayy and Absal leave the more populated island-nation to cultivate philosophy together on the island of Wāqwāq. "I would think that they saw that the solitary man could be more than one," Narbonni writes, "and they could become a group of solitaries, benefiting each other. Although in our time this kind of solitude is impossible without being part of a state . . . I saw the need to include the words of Avempace on the *Regime of Solitude*, to show that the solitary philosopher could also be part of an imperfect state like the states on the edges of climates."³⁹

But Narbonni had a more radical purpose than just flouting a generation-old legal ruling. His pedagogical argument comments on the image of the ladder of teaching, which places theology at one end, philosophy at the other. Ben-Aderet and his supporters from Barcelona sought to anchor theology in the ground; to reach philosophy, the student had to climb each rung, reaching the other end later in life. Ha-Meiri of Perpignan, on the other hand, saw not two ends of a ladder but rather parallel fields of study that should be studied at once. In *Yehiel Ben-'Uriel*, Narbonni radically turns the ladder and plants the philosophical end in the ground, setting physics, astronomy, cosmogony, and metaphysics as the necessary rungs of education before reaching theology. In his view, the practical exploration of nature is a prerequisite for the possibility of communion with God.

Narbonni stressed the role and resources of language for expressing abstract or transcendental ideas. His introduction informs his readers that Ibn-Tufayl conveyed his philosophical views through prose, rather than through analytic writing, to criticize traditional philosophical language that did not allow the possibility of self-taught philosophy. In using prose, Ibn-Tufayl further emphasized the story's privileging of autodidacticism when he "implied that . . . tradition is a great obstacle in the self-directed exploration of the truth." Because concepts are not necessarily dependent upon intellectual tradition, Narbonni asserts that the story gives "a direction to the ways we can practically construct concepts."[40] In summing up, he presents solitude as the optimal social form for the autodidactic life, because it creates an imaginative world, allows the acquisition of fresh impressions, and avoids the cohesive force of tradition. Thus excommunication may lead to exilic epistemology in which autodidacticism represents not merely the only way to study but also the best way to study nature.

In the body of the text, Narbonni again makes reference to pedagogy. In the third part of the story, Ḥayy uses trial and error to discover the basic laws of physics and medicine. In one of the story's most noted episodes, Ḥayy performs a postmortem on the body of the mother-gazelle that nursed him. When he opens her heart he discovers that congealed blood fills the right chamber; he hypothesizes that this was the cause of her death and also that, in fact, this chamber housed the soul that left the body when she died. Narbonni describes the dissection as a practice of inductive reasoning, a process of trial and error, "the way of Ḥayy, always to take evidence from embodied natural evidence and from this to come to concepts."[41]

In addition to promoting the adolescent study of philosophy, Narbonni also attempted to advance autodidacticism as an alternative in the absence of adequate education. As a result, he praised the principle that every man should form concepts inductively, through sensual experience. One who studied alone was independent of the opinions of others, whereas one who learned sitting at the feet of his teachers could be no more than "a domesticated animal led by the bridle." Narbonni prefers "bridle-less boys," as they were described by the opponents of philosophy, since "from embodied interaction come concepts," but in a teacher's presence "there is no difference between a man that is led by transmission and an animal that is led by the bridle; as the poet said, '*Be ye not as the horse, or as the mule, which have no understanding*' (Psalms 32:9)."[42]

Sudden Perception and the Kiss of Death

Narbonni's autodidactic agenda depended upon the direct, unmediated perception of nature, especially as exemplified by sudden perception. As far back as his early writings he counterpoised sudden perception and study as polar opposites.

In his commentaries on Averroes' *Possibility of Conjunction* and al-Ghazzālī's *Aims of Philosophers* he claimed that there are people who can attain knowledge without a teacher. These extraordinary abilities come to light when a person obtains knowledge very quickly, because autodidactic ability is connected to a sudden and immediate acquisition of the truth, a fresh impression, which according to Narbonni reflects the most advanced way of learning. In the introduction to *Yehiel Ben-'Uriel* he again sets gradual study against the notion of sudden perception: "There is knowledge of the divine things that is acquired by transmission and emanates from other knowledge; on the other hand, there is also the sudden perception" that can carry the intellect far beyond natural phenomena into an understanding of the mechanisms of nature.[43]

In *Aims of Philosophers,* al-Ghazzālī writes that acquisition of knowledge without teachers may seem supernatural, but some people need a teacher and others do not. Even so, although every teacher has had a teacher, the chain of teachers could not be infinite, and therefore all knowledge had an independent origin at some point in the distant past and thus could potentially be acquired independently. Al-Ghazzālī refers to "the guarded slate" (*al-lūḥ al-maḥfūṭ*) on which thoughts and concepts exist in potential form before becoming imprinted on the soul. Narbonni picks up on this idea, blending the notions of the guarded slate and the blank slate, and writes that God "scribes concepts on this slate just as he imprinted the law on Moses' tablets."[44]

Narbonni's philosophical approach interplayed with existential concerns. The miseries that defined his life extended beyond a subjective exilic worldview. The king of Majorca passed a wave of ordinances marginalizing Jews, leaving the Provençal communities more and more isolated; and then, beginning in the 1330s, a wave of violence rose up, targeting Jews. The first large-scale anti-Jewish disturbances broke out in Provence-Catalonia in 1331. Later, in 1348, Jews were accused of generating the Black Death, leading the Christian masses to storm the Jewish quarters and kill many people, including judges, prominent scholars, and the majority of the board of elders. Between the plague victims and those Jews who died in the riots, some Provençal-Catalonian communities were almost completely annihilated. Hopelessness about the possibility of reforming society replaced the open-minded worldview that Narbonni had espoused in Perpignan. "And now an enemy appeared and robbery showed up in our windows," he writes in the introduction to *Yehiel Ben-'Uriel,* "those who were destined to the plague and those destined to the sword." Pessimism diminished his faith in cosmic and social order: "As I was observing these catastrophes, I stood puzzled and then I fell on my face and screamed in a loud voice—God has poured his anger on the city of Barcelona ... and chaos came to the world, killing righteous and vicious people together."[45]

Perhaps it comes as no surprise, then, that Narbonni had a fascination with death. The active exploration of nature as delineated in *Yehiel Ben-'Uriel* eventually leads to a communion with God and consequently to an ultimate felicity, sometimes identified with death. He further connected the horrific visions of death to the question of study. Self-directed experimentation and observation aimed to elevate the soul to its highest level of communion with the necessary being, where the soul "would not let off the ecstasy of communion until death ... This is what our sages called *mitat neshika*—'the kiss of death' [*morte osculi*]—a delivery of the soul to God and a peaceful death to the body."[46]

In *Yehiel Ben-'Uriel*, Narbonni supplied the people of Perpignan with additional arguments—for experimental exploration and empiricist deductions, sudden perception, the confines of transmitted knowledge, and spiritual elevation that climaxed in *morte osculi*—to bolster their position that adolescents should be allowed to study philosophy. *Yehiel Ben-'Uriel* echoes metaphors from the controversy, especially the idea that studying with teachers can turn a student into a "domesticated animal led by a bridle."

Back to the Title: Asher Ben-Yehiel and *Yehiel Ben-'Uriel*

Narbonni resided for a short time in Barcelona, the cultural and intellectual antipode of Perpignan. The scholars of Barcelona not only were conservative but also had the cohesive means to enforce their worldview. During the late thirteenth and early fourteenth centuries, the Kings of Aragon dictated a constitution for the Jewish community of Barcelona that empowered the board of *seniores*, all of whom came from wealthy families, to force its authority on their communities through the imposition of fines and punishments, including excommunication and expulsion from the city.

Narbonni's commentary on Ḥayy's story may have inspired the people of Perpignan, but surely, in 1348, the people of Barcelona did not respond to it as well. And though Ben-Aderet passed away in 1310, his followers kept the conservative spirit alive and appointed as his heir another leading scholar of Jewish law, Asher Ben-Yehiel (1250–1327). Ben-Yehiel was not only the most extreme opponent to the study of philosophy but also the one who had urged the issuing of the excommunication ruling.

Born and raised in the Ashkenazi communities of Germany and France, Ben-Yehiel fled in 1303 and wandered through northern Italy and Provence, passing through Perpignan to Barcelona, where Ben-Aderet warmly welcomed him. In Perpignan he witnessed a public speech about philosophy and sarcastically noted that "only a few Jews were left there" still studying and practicing the Jewish law. Ben-Yehiel, however, was not only a passive observer. On his arrival in Barcelona in 1304, he wrote a letter to Aba Mari of Lunel urging him to confront the danger embodied in the teaching of philosophy and offering to summon a meeting of the boards of elders of all communities in Provence-Catalonia "to influence the heart of the people of Israel to strongly hold the revealed laws of Moses" and to keep them away from philosophy.[47] During the 1320s, Ben-Yehiel rose to become the chief rabbi of Toledo, where he infamously applied Jewish law strictly. While consistently opposing Maimonides' philosophy and religious writings, he hardly hesitated to apply his strict interpretations of Jewish law, lightly issuing excommunication rulings and even expulsion from the city.

Ben-Yehiel passed away ten years before Narbonni entered the conservatives' sphere of influence—Barcelona and Toledo. But his followers carried on the austere and strict spirit for decades, disciplining the crowd to stick to the rigorous interpretation of Jewish law and to distant themselves from philosophy. Nevertheless, Narbonni risked clashing with local leadership. He not only expressed his views in favor of philosophy but also commented on the prototypical text concerning the teaching of philosophy to adolescents.

Thus in the mid-1340s Narbonni moved into the territory of his enemies—or, as he put it, into "an imperfect society," a "social exile" that allowed solitary philosophers to gather into groups to sustain their way of life in complete indifference to the happenings in their surroundings. Solitude became a prerequisite for autodidacticism. Liberated from superimposed authorities of either metaphysics or pedagogy, Narbonni argued, a scholar or a group of solitary scholars could freely explore nature and God.[48]

Narbonni's title, *Yehiel Ben-'Uriel,* is not just a play in verse and rhyme. Rather than literally translating Ibn-Tufayl's title (which would render the name Ḥayy as Hayy, Hayyim, or Yihya in Hebrew), he resolved to use the name Yehiel, implicitly addressing the title, and the book itself, to the prominent enemy of Perpignan, Asher Ben-Yehiel. As apparent, he mockingly appended the name Yehiel to the quintessential boy-philosopher, subverting and braking down the authority that presided over both the ideology and the social conditions against which he spoke out. In writing a text that perfectly fit the philosophical taste of Perpignan but stood in firm conflict with local intellectual culture, Narbonni turned the commentary into a political statement.

Despite the solitary nature of his own life, Narbonni's works circulated widely. His many commentaries were used to help teach the various views of medieval philosophy, clarifying problematic places in the primary philosophical texts and making them accessible to those outside of the small philosophical circle trained in the esoteric language of text. His commentary on Ḥayy Ibn-Yaqẓān, in particular, popularized medieval philosophy beyond the social confines of religion, stressing pedagogical restrictions on the study of philosophy. The sheer number of surviving manuscript copies of Yehiel Ben-'Urie stands as testimony to its popularity. A few dozen survived, more than of any of his other works, indicating that his legacy stood largely on the strength of this work and on his commentary on The Guide for the Perplexed.

Regardless of his failure to gain followers and students during his life, his works were sources for Jewish scholars during the early modern period, especially for those interested in criticizing dogmatic philosophy and promoting the new experimental and empirical sciences. One of the first to criticize the Aristotelian physics, Hasdai Cescres (d. 1412), who lived in Barcelona and Saragossa and began writing about philosophy a few years after Narbonni died, argued against the Aristotelian definitions of time and space as finite and offered the possibility of motion in a void that would make time independent from the motion of the heavenly bodies. He closely read the works of Narbonni and echoed Yehiel Ben-'Urie in his rejection of metaphysical principles superimposed on physical exploration, presenting instead the idea of the exploration of nature through practice.[49]

In the seventeenth century, too, Narbonni's ideas played a role in controversies regarding cosmology and scientific practices. Joseph Delmedigo (1591–1655), a student of Galileo and a keen advocate of experimental science and Copernican cosmology, read Narbonni and called it "profound and sublime." Pointing to Narbonni's satiric turn of phrase, Delmedigo wrote, "He spoke in the language of riddles and metaphors."[50] Also a wandering scholar with few social attachments, he followed in Narbonni's footsteps in challenging traditional social and intellectual authorities and argued that adolescents wandering in the world could perceive nature better than old sages.[51] In Amsterdam, Delmedigo, a friend of Spinoza's father and a colleague of his teacher Menashe Ben-Israel, circulated and promoted Narbonni's commentary, which eventually captured Spinoza's attention. This led to Spinoza's involvement in the translation of Ḥayy Ibn-Yaqẓān into Dutch. Just like Narbonni, Spinoza lived in isolation, not subordinated to any religious community. They both had been excommunicated from the Jewish community.

The most important figure in this story's continuation, however, was Jochanan Alemanno. During the 1480s, he keenly read *Yehiel Ben-'Urie* and scribbled extensive marginalia on a manuscript that today sits in a library in Munich.[52] Alemanno turned out to be the key figure in the translation and transmission of *Yehiel Ben-'Urie* into Latin during the turmoil in Florence in the early 1490s.

CHAPTER THREE

Defying Authority, Denying Predestination, and Conquering Nature

Florence, 1493

During the fifteenth century, manuscripts of Narbonni's commentary on *Ḥayy Ibn-Yaqẓān* circulated among Jewish communities from Barcelona to Provence, finally making their way to Italy. In the early 1490s, Giovanni Pico della Mirandola encountered one of these manuscripts, and in 1493 he had it translated into Latin. The manuscript, a straightforward translation with little annotation or marginalia that now lies in the University of Genoa library, gives no indication of the context that drew Pico to *Ḥayy Ibn-Yaqẓān*, but social occurrences in the backdrop supply some leads.[1] Pico's interest in the story cropped up during the years 1492 to 1494, when Florence went through a period of cultural turbulence. Controversies over astrology, sodomy, and the arguably promiscuous street activity of boys split the city and stimulated heated disputes regarding the human capacity to shape destiny. Pico took a decisive stance. He set his mind to refute astrology and hectically worked to complete his work, *Disputationes adversus astrologiam divinatricem*. At the same time, he personally experienced the cultural uproar, until he was poisoned in November 1494.[2] The Latin translation of *Ḥayy Ibn-Yaqẓān* supported Pico's argument against astrology, presenting autodidacticism not only as independent education but also as a form of liberation, freeing man from the chains of the cosmos and allowing him to explore and thus to conquer nature. Such a radical stance came into being not in pure contemplation but rather through internalized personal experiences.

Pico was new to Florence. After Pope Innocent VIII condemned his works, the tall, long-haired, twenty-five-year-old Pico arrived in Florence in 1488 and came under the protection of Lorenzo de' Medici—Lorenzo the Magnificent—ruler of the city. He joined the poets Girolamo Benivieni and Angelo Poliziano, the Platonist Marcilio Ficino, the artists Sandro Botticelli and Domenico Ghirlandaio, and other poets, philosophers, musicians, and artists who celebrated the rediscovery of classical literature, the incorporation of nature into painting, and the notion of Platonic love—the longing for ultimate happiness through spiritual coupling with friends, nature, truth, and God.

Later, in 1490, Pico convinced Lorenzo to invite Girolamo Savonarola, an ascetic Dominican priest with a strong faith in predestination, to come under his patronage and to join his circle of scholars.[3] Lorenzo did not attract such celebrated philosophers and artists simply because of his disposition toward art and philosophy. Controversial humanists, some of whom got into trouble with the Roman church, turned to Lorenzo's court, shaping it into a revolutionary intellectual center that challenged the traditional authority of Paris and Rome. With its absolutist Medici dynasty and cultural defiance against Rome, Florence stood as a political island surrounded by enemies, from France to the north, the Republic of Venice to the west, and Rome and Naples to the south. The political challenges eventually materialized between 1492 and 1494, after Lorenzo died, when Medici rule and culture collapsed and the circle of philosophers and artists lost its protection. Those were also the last two years of Pico's life. He spent them enclosed in Villa Fiesole, some miles north of Florence, where he thirstily read texts that facilitated the refutation of astrology—including the text about the child prodigy who acquired habits and knowledge through firsthand experience.

Pico extensively used the pattern of the child prodigy as best exemplifying the liberation of man from predestination and tradition, implicitly referring to his own childhood experience. Born on February 24, 1463, in a small city-state in the plain of Emilia, the youngest of five children, and trained from an early age in Latin and possibly Greek, Pico was considered a precocious child with an amazing memory. His father, Gianfrancesco Pico della Mirandola, count of Mirandola and Concordia, died in 1467 when his son was only four years old, after which his mother, Giulia, raised him. Without the traditional authority of a father, Pico quickly developed a self-reliant approach to learning and a sense of entitlement regarding his right to know everything and to defy authority.

Pico's mother had huge plans for her youngest son. She envisioned him as a prince of the church and put him up for the most coveted ecclesiastical position, the cardinalate, from which the pope himself is selected. In 1477, at the age of fourteen, the young count enrolled at the University of Bologna to study canon law, a step his mother thought necessary for his ascension. After her death in 1478, however, Pico came to see the world from the vantage point of an aristocratic orphan and began moving restlessly from one university to another, refusing to accept any intellectual authority. In 1479 he enrolled in the University of Ferrara, where he participated in his first public disputation with Leonardo Nogarola on the subject of the immortality of the soul.[4] Only sixteen years old, Pico was praised by scholars for his resourcefulness and genius. Tito Vespasiano Strozzi, one of his masters in Ferrara, dedicated a poem to him in which he writes, "You were luckily endowed with in-born talent and are brilliant in all learning / To whom has it been more granted to know the innumerable causes and vicissitudes of things and the foundational laws of nature / Who measures with such great judgment the orbits of sun and moon and the bright stars of heaven." In addition, Strozzi concludes, "you have the great gift of good fortune, to which goods add also physical gifts."[5]

Pico continued to leave exceptional marks on the various intellectual communities. From Ferrara and its tradition of public disputations, he moved in 1480 to the University of Padua, where he preferred extracurricular circles over the formal and authoritarian classes of the university. Over the next four years the wild prodigy restlessly wandered among different humanist centers, including Florence—in 1484—and Paris, and he set his mind to reinventing philosophy. He rejected the scholastic unquestioned philosophical principles and listed nine hundred true theses, from ancient theology to medieval philosophy, which were published in Rome (1486) under the title, *Conclusiones philosophicae, cabalasticae et theologicae*. He not only challenged the hegemonic philosophical tradition but also issued a general invitation to scholars, who ridiculed his sense of entitlement, asking them to face him in a public dispute and offering to pay their travel expenses.

The opening speech he planned for the occasion, *Oration on the Dignity of Man*, took on the critics who scorned his arguments because of his own youth and declared that he had decided "at a mere age of twenty-three years . . . [to] dare to propose a discussion on the sublime mysteries . . . in a famous city, before a vast assembly of learned men, in the presence of the apostolic senate."[6] Although he knew that he would be judged immodest, holding an excessively high opinion of himself, as well as reckless and imprudent, he carried on, "so you see the difficulties into which I have run . . . and I can find comfort in what Timothy was told: 'Let no man despise thy young age [*Nemo contemnat adolescentiam tuam*].'" Despite all the adversaries lined up against him, Pico still hoped that the conclusions drawn from his *Theses* "will be reached with less regard to my age than to the ultimate outcome of the contest."[7]

In the end, though, it was not his young age that prevented the public debate. It was Pope Innocent VIII, who halted it and ordered a committee to review the orthodoxy of Pico's *Theses*. Thirteen of Pico's nine hundred theses were condemned; one, for example, claimed that the best science by which to prove the resurrection of Christ is natural magic and Kabbalah. In response, Pico wrote an

Apologia that sought to defend his *Theses*, leading to the condemnation of the entire work and to his persecution at the hands of the Roman Catholic Church. Hoping to escape, he fled to France in 1488, where he was arrested and imprisoned at the demand of papal nuncios. Several Italian princes—all prodded by Lorenzo de' Medici—intervened, leading King Charles VIII to release him. The pope was persuaded to allow Pico to move to Florence, where he would live under Lorenzo's protection. The bold young scholar, who had challenged the authority of the church, now found himself in Florence—a city that took as its symbol the biblical David, who defied and defeated Goliath.

Although Pico was the youngest in the Florentine humanist circle around Lorenzo, his colleagues regarded him as a self-taught philosopher, naturally endowed with genius. "Nature seemed to have showered on this man, or hero, all gifts," Poliziano wrote. "He was tall and finely molded and from his face divinity shone fourth. Pico is gifted with a prodigious memory, in his studies he was indefatigable and in his style he was perspicuous . . . He stood high above the reach of praise."[8] In February 1491, Lucio Phosphorus, bishop of Segni, wrote to Poliziano that Pico is "the one person in whom Nature seems to have amassed all its gifts and to have exercised all its powers."[9] Even Niccolò Machiavelli, in his *History of Florence,* in retrospective gives a sense of the public image that Pico held in Florence. While eulogizing Lorenzo de' Medici, who died in 1492, Machiavelli noted that "Count Giovanni della Mirandola, a man of almost supernatural genius, after visiting every court of Europe, induced by the munificence of Lorenzo, established his abode at Florence."[10]

It was not only a circle of friends who promoted his reputation as self-taught philosopher. Over the years, Pico has been portrayed in various, often conflicting, ways—as a liberal, a magus, an obsessive reader, and a mystic[11]—but all accounts commonly relied on his self-valued image as an autodidact. Paul Delaroche painted Pico and his mother (1848), echoing Renaissance images of Christ and the Virgin. In his *Oration* Pico explicitly proclaimed his autodidactic vision, paraphrasing Horace's famous autodidactic claim, *nullius addictus judicare in verba magestri*—"not compelled to swear to any master's words"—and declaring that he would not take for granted the authority of traditional philosophy or trust any philosopher's words. Instead, he had resolved himself "to range through all masters of philosophy, to examine all books, and to become acquainted with all schools."[12]

Personal experience was internalized into philosophical worldview, producing a distinctive stance that rejected authority and celebrated self-molding. But Pico's core stance was also subjected to social pressures, engendering various

Pico on the lap of his mother, who is holding a book. The image replicates early modern images of Christ, but instead of embodying messianic revelation, here Pico quintessentially represents the Enlightenment plea for autodidacticism. Paul Delaroche, *The Infancy of Pico* (1842). Musée des Beaux Arts, Nantes, France.

degrees and meanings of autodidacticism. In viewing Pico's interest in the story of *Ḥayy Ibn-Yaqẓān* as embedded in the larger sociopolitical crisis that Florence experienced in the 1490s, new directions come to light, pointing at particular motives that prompted him to radicalize his arguments in favor of autodidacticism and to expand its meanings.

Florentine culture during the 1470s and 1480s added new dimensions to the question of knowledge acquisition. Rather than the medieval view that perceived passive contemplation of transcendental God at the heart of any process of learning, Florentine scholars and artists promoted the notion of Platonic love, stressing that knowledge is acquired through a process in which man spiritually couples with objects, people, nature, and God.

With these new self-indulging meanings, Renaissance culture reached its climax in Florence in the early 1490s. Notions of spiritual coupling and interactions worked into practices across various layers of society: brotherhoods promoted the exchange of ideas, painters celebrated human bodies and placed them in the center of nature, scholars stressed the coupling with ensouled nature, and men of all social layers rejoiced in sexual coupling with other men and swept boys into homosexuality.

The death of Lorenzo the Magnificent in 1492, and various acclaimed supernatural omens that accompanied it, suggest a political transformation—the fall of the Medici and their artistic and scholarly circles. New voices that disapproved of the Medici culture particularly hung onto the practice of sodomy as its quintessential manifestation. Priests preached against the "sin city," where boys engaged in sodomy and astrologers predicted that Florence would be turned upside down. The social unrest engendered a call to save the city through a religious and educational transformation, one based on doctrinal conformity that would be superimposed through institutional discipline. The ascetic Dominican priest, Savonarola, gave the tone to the social unrest and stimulated a "cultural revolution" based on pious contemplation of God.

Such developments threatened both the ideas and the practices of the humanist circle. Its members formed and promoted the notion of coupling as a process of knowledge acquisition and projected it onto various cultural customs, particularly to sexual practices involving sodomy of boys. Pico resolved to refute astrology, using various examples of children who were not predestinated by the influence of the stars but rather acquired habits and knowledge through firsthand experience. He then turned to the treatise of *Ḥayy Ibn-Yaqẓān* as a philosophical exemplar of how anybody, even a boy on a desert island, could become a philosopher and take charge of his destiny. While early in his career

Pico staged self-teaching as a rejection of authority and tradition, after the social crisis of 1492 he presented autodidacticism as evidence disproving predestination and determinism.

In dismissing the spiritual qualities of nature, Pico liberated man from astrology and articulated a new relationship between man and nature, positing a dead and passive nature that could be experimented upon, manipulated, conquered, and controlled. He recast his early humanist views into a particular natural philosophy that fostered experimentalism. Social pressures of the time combined an urgent need to refute astrology with personal experience in autodidacticism, stimulating Pico's interest in Ḥayy Ibn-Yaqẓān.

A Plain Manuscript and the Enigmatic Translator

The manuscript-translation of *Ḥayy Ibn-Yaqẓān* in the library of the University of Genoa contains no autobiographical insertions that could illuminate Pico's motives, the date, or even the place of engagement with the Andalusian philosophical tale.[13] Pico reveals his familiarity with *Ḥayy Ibn-Yaqẓān* only in an autobiographical reference in the first book of the *Disputationes adversus astrologiam divinatricem*, where he brings examples of anti-astrological writings.[14] In particular, he conjures up Ibn-Tufayl to demonstrate the confusion between philosophers and astronomers, on the one hand, and astrologers, on the other: "Someone may make the error that, since the names of certain philosophers and astrologers are similar, he may think that the same people have written both philosophical and astronomical works ... There is an Abubater who wrote on nativities, and another man with the same name wrote philosophical works, most notably the book *How Anyone Can Become a Philosopher on His Own*, which last year I translated from Hebrew into Latin. But the former is the son of Alchasibi, the latter, of Tofail."[15] The forthright title he gave the translation indicates some of Pico's arguments against determinism and astrology. The title not only praises self-directed learning, it also points out that through experience one has the freedom to mold his destiny, sometimes in intricate and unpredictable ways.

Pico's encounter with the treatise was part of a larger interest in oriental studies. He took an interest in Hebrew, Aramaic (Chaldean, as it was then called), and Arabic philosophical and mystical texts. He first encountered "oriental philosophy," as he often called it, very early in his education at the University of Ferrara.[16] His engagement with Hebrew and Arabic started as an extracurricular initiative that intensified after he met a Cretan Jew, Elijah Delmedigo (1450–92), while studying at the University of Padua during the two-year span from 1480 to

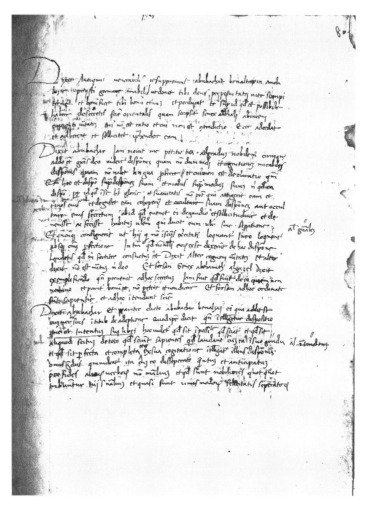

Marginalia on the first page of Pico's Latin manuscript-translation of Ḥayy Ibn-Yaqẓān. The insertions suggest other ways of translating philosphical notions. The handwriting in the marginalia and the text are identical, indicating that the surviving copy is not the one Pico used but an extra, made by a copier from Pico's original. Biblioteca Universitaria di Genova, cod. A, IX, MS 29, fols. 79v–116r. Courtesy of Biblioteca Universitaria di Genova and Ministero per i Beni e le Attività Culturali.

1482.¹⁷ Although Pico associated with various Jewish scholars who might have informed him about the existence of *Ḥayy Ibn-Yaqẓān*,¹⁸ he came to know the work only after his arrival in Florence in 1488. During his first visit to Florence in 1484, he met another Jewish scholar, Jochanan Alemanno, a Florentine scholar of Kabbalah and biblical exegesis. The two only started working together, however, in 1488. Alemanno, who seemingly introduced Pico to the treatise, cultivated Kabbalist commentaries on the Song of Songs, which literally describes an erotic love affair between a man and a woman, but for Alemanno it stood as an allegory for the coupling of the human soul with God.¹⁹ He later composed *A Gate of Desire* (*Sha'ar ha-ḥeshek*), in which he describes King Solomon as an ideal human archetype, perfectly moral and endowed with complete knowledge about nature.²⁰

Alemanno also composed a work that echoes Hayy's story and presents a program for self-directed learning. The book, titled *Ḥayy ha-'olamim*, describes the stages of human development—physical, emotional, social, and spiritual—from the moment of conception to the attainment of divine knowledge, the ultimate perfection. He described the work as "a book on the history of the righteous man" that presents a program for molding boys from early infancy through gradual learning, a process that would allow anyone who follows its autodidactic program to reach spiritual climax.²¹

In addition to reconstituting ideas from *Ḥayy Ibn-Yaqẓān*, Alemanno set in motion the circulation in Italy of the Hebrew manuscript of Narbonni's commentary on the treatise. A Hebrew copy of *Ḥayy Ibn-Yaqẓān* with Narbonni's commentary lies in the Bavarian State Library. Full of illustrations of men and animals, it also contains a commentary in the margins (or supercommentary) by Jochanan Alemanno, who commented only on select parts of the text, on burning issues of the day that engaged Pico and others in Florence. For instance, on the question of spontaneous generation, which represented the ideal case of liberation from predestination, Alemanno describes man's creation as a purely physical and material process²² and Hayy's climactic coupling with the "necessary being" (God) as climbing on the ladder of the ten Kabbalist spheres.²³

The connection between Pico and Alemanno was not merely on the level of sharing ideas and sources. In 1489, while exploring the Kabbalist exegesis on the Song of Songs with Alemanno, Pico wrote and published *Heptaplus*, a sevenfold commentary on the first twenty-seven verses of Genesis. Alemanno's mark can be seen on the work, especially in the choice of exegetical genre and the occupation with *gematria*, the Kabbalist method of breaking down words to numbers and then composing new words and meanings out of them. More important,

Diagram of the ten Kabbalist spheres, which Alemanno inserted into Narbonni's *Yehiel Ben-'Uriel*, next to the text describing the way Ḥayy achieves communion with God. The insertion illustrates Alemanno's belief that the secrets of nature can be unfolded through the wisdom of Kabbalah. Later on, Pico's translation follows the same line of argument, presenting the moment of Ḥayy's communion with God as a Kabbalist revelation. Courtesy of Bayerische Staatsbibliothek, Munich, cod. Hebr. 59.

however, the preoccupation with the Genesis description of the creation allowed Pico to use Alemanno to further explore his early interest in the idea of the first man as the origin of philosophical debate.

In the *Oration* Pico introduced his view that the first man's freedom from predestination made him the supreme creature in nature. He writes that God took man, "this creature of indeterminate image," set him in the middle of the world, and told him, "We have given you, Adam, no fixed seat nor features proper to yourself nor endowment peculiar to you alone, in order that whatever seat, whatever features, whatever endowment you may responsibly desire ... constrained by no limits, you may determine it for yourself, according to your own free will, in whose hand we have placed you ... I have made you ... as the free and extraordinary shaper of yourself. Fashion yourself in the form you will prefer."[24] Here, Pico actually reshapes Genesis 1:28: "Be fruitful, and multiply, and replenish the earth, and subdue it: and have dominion over the fish of the sea, and over the fowl of the air, and over every living thing that moveth upon the earth." Rather than highlighting physical reproduction, as the Bible seems to do, Pico emphasizes man's autodidactic possibilities, thus interpreting God's first imperative to humanity—to reproduce and conquer the earth through the active exploration of nature. The coupling with nature, for Pico and other Kabbalists, became a high form of worship of God.

With the assistance of Alemanno, the figure of the first prehistoric man, or Adamic man, and his godly vocation to conquer nature was further developed in the *Heptaplus*, in which Pico talks about the place of self-knowledge and self-directed learning in fulfilling humanity's need to "enter into our very selves" in order to study and understand things quite remote from themselves.[25] It is not only the angels, he asserts, who can perceive the whole of nature: "Any intelligent creature ... includes all things in himself in some degree when, filled with their forms and reasons, he knows them." Felicity, for Pico, is present in the instant of knowing, the sensation of first imprints on the consciousness, when the naming of things "returns nature to its beginning." He then first makes passing reference to the writings of Ibn-Tufayl, who seems "to speak about natural happiness."[26]

Clearly, then, Pico knew of Ibn-Tufayl's autodidactic philosophy at least as early as 1489, while writing the *Heptaplus*, which suggests that Alemanno was the prime vehicle for introducing Ḥayy's story to the humanist atmosphere of Florence and to Pico in particular. But that does not prove that Alemanno acted as Pico's translator, especially since the men had no contact during 1492 and 1493, when the Latin translation appeared.[27]

Franco Bacchelli argues that a monogram of Christ in the manuscript indicates that a Jew who converted to Christianity acted as translator. He names a Sephardic Jew, Isaac Abramo, as the most likely suspect because, in 1492, Abramo worked for Pico in Florence, probably assisting him with some Hebrew texts.[28] Abramo's draft of the translation of *Ḥayy Ibn-Yaqẓān* has not survived, but a neater copy was found among the papers belonging to the heirs of Pierleone da Spoleto, Lorenzo de' Medici's physician, and ended up in Florence in Roselli del Turco's house. Bacchelli concludes that Pierleone made the copy sometime in 1492, before his death.[29] Thus Pico engaged with the story of Ḥayy when he moved from merely centering man in the cosmos to freeing him from predestination. The process that took place in Florence also paralleled his increasing loathing of astrology.

From Neutrality to Fervent Opposition to Astrology

Early in his career Pico displayed a faint curiosity about astrology. Darrell Rutkin has recently found that in some early poems Pico used figurative astrological language,[30] and he had his horoscope drawn up by his close friend Girolamo Benivieni. Later, in the late 1480s, he also incorporated some astrological references into his writings.[31] Mostly, however, Pico concentrated on promoting independent exploration of nature through the Kabbalah and natural magic. In the *Oration* he distinguishes between good and bad magic. Bad magic depends entirely on the work and powers of demons "and is, in my faith, an execrable and monstrous thing." Good magic, however, represents "nothing but the absolute perfection of natural philosophy." As a result, good magic acts as nature's servant rather than its artifact and works by locating and activating the virtuous powers that have been scattered by God and hidden in the world. The magus—in Persian, "the expert in divine things"—is one who uses a deep knowledge of the secrets of nature to "[marry] earth to heaven, that is, the lower things to the higher endowments and powers." Thus the "natural magic I am discussing [is] more ardently aroused to the worship and love of their maker."[32] The power of *naturalis magia* brings its practitioner to a communion with God through a practical exploration of nature's interconnected structure.

Never printed in Pico's lifetime, the *Oration* gained enormous fame only after his death. But Pico did publish *Conclusiones*, in which he traced universal wisdom back through time—from scholasticism all the way back through the Arabs to the Greeks and then to the ancient sages of *prisca theologia* (ancient theology),

Zoroaster, Moses, Hermes, and Pythagoras—presenting a syncretism that emphasized the importance of natural magic.

"*Naturalis magia*," he writes in the *Conclusiones*, "is the practical part of natural philosophy (*practica naturalis philosophiae*)," of which, through Kabbalah, one could "practice every formal quantity, continuous and discrete."[33] All knowledge of nature's deepest secrets should be explored using the methodology of natural magic, or as he called it, "practical natural philosophy," which begins below in practical interactions with the physical world and rises to the knowledge of divinity.[34]

Even the papal condemnation of his *Conclusiones* did not deter Pico from following this line of thinking. Later, in the *Apologia*, he continued to stress the connection between natural magic, Kabbalah, experimentalism, and autodidacticism.[35] The word *magus*, Pico explains to his ecclesiastical readers, does not refer to a supernatural practitioner but rather is the Persian name for a philosopher. Thus "*naturalis magia* . . . is properly the practical part of *scientia naturalis*, which presupposes an exact and absolute understanding of all natural things."[36] It seems clear that in his early works he extensively discussed natural magic and Kabbalah, but he rarely mentioned astrology. His silence about astrology is all the more surprising given that its predictions played such an influential role in his world.[37]

Pico finally broke his silence about astrology after his arrival in Florence. In the *Heptaplus* he only implied his reservations about astrology, by reducing the influence of the heavenly bodies to mere physics. He describes man's direct connection with the cosmos through light (*lux*): "Every virtue of the heavens is conveyed to earth by the vehicle of light."[38] Physical light, not spiritual influences, connects the heavens with the earth, and thus dead nature can best be explored through natural magic—that is, practical natural philosophy. At some point, Pico's attitude toward astrology changed, moving from implicit reservations to an active mocking of astrologers, which after 1492 turned into a vehement attack against the whole discipline of astrology.[39] Rather than merely evolving out of philosophical considerations, however, Pico was reacting to radical changes in his surroundings.

Urgent Times and Burning Passion

In the late 1480s Lorenzo de' Medici gave Pico a villa in Fiesole, where Pico worked on the *Disputationes* from 1492 until his death, on November 17th, 1494.[40] Eugenio Garin notes a Greek epigram by Poliziano bemoaning the fact that Pico's burning obsession to confront astrology kept him holed up in his suburban villa, "working on refuting those *charlatans* rather than spending time with me in the city."[41] Pico himself stated the purpose of the work in the preface, where he writes that "people of our time of all ages and sexes are prone to this kind of superstition, therefore I have to fight the charlatans and show the people the truth. And I don't care about the opinion of the plebes because I do not seek popularity but only the light of the truth and public good."[42]

Based on the papers he left behind, it appears that Pico hectically worked on the *Disputationes* until the time of his death. His nephew, Gianfrancesco, testifies that he found among his uncle's notes "fragmentary and incomplete, messy, badly written, not yet developed, twelve books against judicial astrology."[43] Gianfrancesco and Pico's editor, Giovanni Mainardi, edited the work for publication; subsequently, Savonarola further edited it, a process that lasted until the final document was published in Bologna in 1496.[44]

But the heavy editing that Pico's drafts required suggests that something happened to Pico during that time to force him into virtual seclusion in his villa in Fiesole, working feverishly on the *Disputationes* but unable to edit his drafts. These drafts or Savonarola's editing of them most probably censored the identities of "those charlatans" and also softened the figurative descriptions of sexual perversity, which in 1492 came to be considered improper.

The Sky Falls on Florence

A strange coincidence of deaths increased the popular belief in astrology. Lorenzo de' Medici died on April 8, 1492. The same night, the body of his physician, Pierleone da Spoleto, was found in a water well. In a letter to a friend, Poliziano recounts the events he witnessed while sitting by Lorenzo's sickbed. By April 1492 Lorenzo had been sick for some months. As his condition deteriorated, various doctors were called to treat him and to experiment with any possible remedy, from a powder of ground pearls to practices of astrological medicine. Pierleone da Spoleto, whose library reflects his expertise in medical astrological practices, came to Florence to offer his assistance.[45] One of his treatments, however, caused Lorenzo to fall into a fever that presaged the end. Understanding his own condition, Lorenzo asked Poliziano "very amiably what your friend Pico was up to." Poliziano replied that Pico stayed in town, worried that he should be something of a nuisance to Lorenzo. "But," Lorenzo replied to Poliziano, "if I myself in turn did not worry that the journey would be annoying for him [Pico], I would very much like to see and speak to him one last time, before I leave you once and for all." "Do you want him [Pico] to be sent for?" Poliziano asked. "Yes," Lorenzo firmly answered, "as soon as possible."[46]

When Pico arrived, Lorenzo was excited to see him. With Poliziano listening close by, Lorenzo told Pico that he "attributes to his love and goodwill toward him the fact that he [Lorenzo] was prepared to give up the ghost more willingly if first he would sate his dying eyes on the sight of a very dear friend." Poliziano then witnessed as the dying man "introduced witty and intimate topics of conversation, as was his wont." As Pico left the room, Girolamo Savonarola entered and exhorted Lorenzo to keep faith. "Nothing, in fact," Lorenzo replied to Savonarola "would be more pleasant should this be in God's plan." Lorenzo soon died; thereafter, the sky seemed to fall on Florence. Poliziano continues his letter:

> The following omens anticipated his death.... Three days before Lorenzo gave up the ghost, some woman, as she was listening to someone preaching from the pulpit in the church of Maria Novella, suddenly, in the middle of a thick crowd of people, leapt up, frightened and agitated, and amid frantic running and terrifying shouts cried, "Look! Look! Citizens! Don't you see the raging bull that knocks this massive temple to the ground with his flaming horns?" Then early that night, as the sky grew unexpectedly dark, the main cathedral's own lantern, which rises above the dome which through its wondrous engineering is unique in all the world, was suddenly struck by lightning, with the result that several vast chunks were toppled, and gigantic marbles—especially toward the side from which the Medici palace is visible—were twisted out of shape by some awesome power and force. In this, the following too was not without future significance, for a single gilded ball, like others visible on the same lantern, was knocked off by lightning, to insure that, given this particular sign as well, harm specific to this family was portended.

Poliziano goes on to describe other celestial events occurring in Fiesole that had astrological implications. "On the night of Lorenzo's passing," Poliziano writes, "a star brighter and larger than usual, hanging over the suburban villa in which he was breathing his last, was seen to fall and be extinguished at the very moment" Lorenzo gave his last breath. "Furthermore," testifies Poliziano, "three nights in a row, torches were reported to have raced down from the hills around Fiesole, all night long, and to have flickered for a while and then vanished over the sanctuary where the mortal remains of the Medici family are interred." As often happens during such radical cosmic changes, animals in the local zoo sensed it first, and "a most noble pair of lions too, in the very cage in which they are kept on public display, so ferociously attacked one another that one was badly hurt, and the other was even put to death. In Arezzo too, right over the castle, twin flames are said to have burned like the Dioscuri [the mythological twins Castor and Pollux who were transformed into the constellation Gemini]. And a she-wolf, just outside the city walls, repeatedly emitted terrifying howls."

Poliziano also included the mysterious death of Pierleone as a sign: "Some people even gave a portentous interpretation to the fact that the most excellent medical doctor of our day, when his method and his prognostications failed him, gave in to despair and, by his self-slaughter, performed a relative's sacrifice . . . to the very prince of the medical family."[47] Pierleone's death generated a wave of rumors. In a letter dated April 14th, the artist Bartolomeo Dei wrote to his uncle about "the strange death of master Leoni who was betrayed by his false science, who some say was mixed with black magic."[48] Some even said that Pierleone da Spoleto applied his astrological practices not only to Lorenzo but also to himself and had, in January 1492, predicted his own death in April.[49]

Taken together, these events created the general impression that catastrophe had struck. In his *History of Florence*, Machiavelli described the political implications of the cosmic-cultural turmoil that befell Florence after Lorenzo died: "As from his death the greatest devastation would shortly ensue, the heavens gave many evident tokens of its approach . . . and hence, soon after the death of Lorenzo, those evil plants began to germinate, which in a little time ruined Italy and continue to keep her in desolation."[50] Such astrological predictions agitated Pico, as book 3 of the *Disputationes* attests; there, he fulminates against political predictions as "totally insane" and writes that "the destruction of cities comes not as a result of a prediction made by stupid charlatans but as a result of the will of God."[51] He had a personal reason to object to such predictions: according to the astrologers, Florence drew its punishment because of the lustful sins concerning boys, sins in which Pico and his circle participated.

Swiping the Boy's Hat

David, the Old Testament youth who stood up to the mighty Goliath, had long been the symbol of the Florentine Republic. Under Lorenzo de' Medici, the city came to identify with the figure of David more strongly than ever. Donatello's *David* stood as the centerpiece in the first courtyard of the Palazzo Medici during Lorenzo and Clarice Orsini's wedding festivities in 1469. Unlike other fifteenth-century sculptures of David, however, Donatello's bronze stands naked with a hat on his head. The statue speaks to more than simply the developments in Italian art; it gets at the heart of fifteenth-century notions of male identity. Whereas girls became brides in their teens, men did not generally marry until their mid-twenties or later. Adolescence thus extended from childhood to full adult maturity for the male in Renaissance Florence, a state that was reflected in the art of the time. Portraits of youthful males have an effeminate quality—slim proportions, long hair, and detached glance—all qualities seen in Donatello's *David*. (For example, the only known portrait of the adult Pico he is portrayed as an effeminate figure.) Significantly, the training and socialization of young men openly depended upon these fifteenth-century Florentine depictions of adolescent males. Rather than proclaiming liberation from social strictures, the frequent homoeroticization of the young male subject reproduced and enforced patriarchal authority.[52]

In an effort to abate the threats represented by obstreperous youth, the adult community instituted social mechanisms for taming the independent wills of their sons. Adolescent confraternities and informal secular brigades channeled the energies of young men into the city's ritual life; ecclesiastical schools and private humanist tutors instilled Christian values and a deep respect for elders. The confraternities led carnivals, and their members wrote and delivered public speeches. Adolescents, at the center of public life and public space, represented the vitality of social organizations.[53] But their exposure to sexual conduct existed in inverse proportion to their centrality in street culture.

In exploring legal documents in Florentine archives, Michael Rocke has found that sodomy occurred throughout the city, involving men from all walks of life. Although recorded fundamentally through its prosecution, its very public nature emerges clearly. Defined as "unmentionable" by social critics and theologians, sodomy nonetheless possessed an irrepressible life of its own. Practiced widely and publicly by Florentines, it produced particular categories of masculinity, structuring "homosocial" relations. In an overwhelming majority of cases, such relations involved an older, active partner and an adolescent, passive partner. Rocke describes in detail the "game" of hat stealing, which he calls "a sort of ritual extortion for sex." Men seeking sodomitical sex would swipe the hats of the boys who attracted them and refuse to hand the caps back unless their advances were returned. "I won't ever give it back to you unless you service me," cried Piero d'Antonio Rucellai, according to his victim, the fifteen-year-old Carlo di Guglielmo Cortigiani, testifying in a trial of 1469. After submitting to sex in an alley, Cortigiani's hat was returned to him.[54]

Certain features of the intellectual and cultural milieu fostered by the Medici

The Confirmation of the Rule, Sassetti chapel, Basilica di Santa Trìnita, Florence. In this fresco, Domenico Ghirlandaio counterpoises the two contradicting cultures of Florence—the pious high culture versus the common street culture. In the upper scene, St. Francis is being received by Pope Honorius III. The scene is set in Florence rather than Rome, the background showing Piazza della Signiora and Palazzo Vecchio alluding to the power and status that Florence had assumed; in humanist circles it was considered a new Rome or Jerusalem. Below, another scene unfolds. Poliziano, surrounded by Lorenzo's boys, seems to plead for mercy. One of the boys straightens his eyes to the painter in a gesture of disclosing a secret. Courtesy of Basilica di Santa Trìnita, Florence.

also favored a greater acceptance of homosexuality: the prominence of beautiful adolescents, the revival of the depiction of the male nude in art, and especially Neoplatonism, with its idealized homoerotic ethos. Leonardo da Vinci, for one, was charged in 1476 with having sodomized seventeen-year-old Jacopo Saltarelli, an apprentice goldsmith and part-time prostitute.[55] Lorenzo counted among his closest friends and companions men with known or suspected homoerotic inclinations. Some even made it into the court records because of sodomy charges.

Angelo Poliziano, Pico's closest friend and author of amorous letters to youths and a misogynous defense of homoerotic love in his drama *L'Orfeo*, found himself implicated in court for sodomy in 1492, and in 1494 a boy claimed that Poliziano had sodomized him not long before he died.[56]

With Lorenzo's death, the comparatively permissive climate that had briefly allowed homosexual activity to flourish came to an abrupt end. A cultural struggle over homosexuality marked the two years between the death of Lorenzo (1492) and the collapse of the Medici regime (1494), while Pico was composing the *Disputationes*. In a letter dated April 7, 1492, only one day before Lorenzo's death, twenty-five-year-old Niccolò di Braccio Guicciardini wrote to his eminent relative, Piero Guicciardini, expressing his fear, shared with his close circle of friends, of the apocalyptic preaching against sodomites that had recently gained prominence in the city's pulpits. Mentioning the evil omen seen in the lightning that had struck the cathedral roof two days earlier, he wondered fearfully whether it was an apocalyptic sign of a forthcoming punishment for the sin of sodomy. According to the nervous young man, the preachers were claiming that "God sent this scourge so that we would repent of our sins, especially sodomy, which he wants to be done away with . . . [Otherwise] the streets will run with blood . . . such that all of us are frightened, especially me. May God help us."[57]

Immediately following Lorenzo's death, the regime indeed began to reinforce the apparatus for controlling sodomy. The Eight of Watch, the city's most important criminal court, initiated some spectacular police actions against adolescents and sodomites and appropriated more extensive powers. Niccolò Guicciardini's worried letter to his relative furnishes a rare private account of the magistracy's energetic intervention and of the growing climate of terror and repression that gripped the city. On April 3, as Lorenzo lay dying, the police rounded up some twenty adolescents, "all of good families," as Niccolò noted, "and interrogated them about their homosexual relations." One promiscuous boy named Duccio "Mencino" implicated none other than Angelo Poliziano. "On the following evening," the adolescent Niccolò wrote to his uncle, "the police sent their retainers to sweep the city's taverns and detain anyone found in the company of a boy. They ordered taverns and innkeepers to deny entrance to any youth or boy who might commit sodomy."[58] If the prevailing acts of sodomy would not end, the preachers pronounced, Florence "would be turned upside down" and come to the same devastating end as the biblical Sodom. The many convictions for sodomy after Lorenzo's death reflected a new cultural climate that stood in opposition to the sentiment felt by Pico's close circle, a circle targeted as the cultural focus of these activities.[59]

Neoplatonic and Erotic Loves

Many members of Pico's circle, in addition to Poliziano, practiced homosexuality and praised homoeroticism in their writings. Some evidence, albeit muted, exists that Pico not only favored looser attitudes toward homosexuality and associated with friends involved in sodomy and homosexuality but that he himself engaged in homosexual relations and sodomizing of adolescents. In the *Disputationes* he tells the story of his friend who had been caught up in sadomasochistic sexual relationships since childhood. Pico explains that his friend was unable to enjoy sexual intercourse unless flogged viciously with a rod soaked in vinegar. One day Pico asked him how he could be so perverse as to desire pain. The friend replied that he had been raised with others who traded sexual favors for the right to hurt those they gratified; the association between pleasure and pain became ingrained in him and never departed.[60] The implication is clear: if even the strangest of human characteristics can be acquired, it stands to reason that, as in Ḥayy's circumstance, more normal attributes can also arise from life's circumstances.

Pico also used common euphemisms indicating the extent of his engagement with homosexuality, and in his 1486 commentary on Girolamo Benivieni's *Dell'amore celeste e divino* insinuated his own sexual preferences and pointed obliquely toward his partner.[61] He praised Benivieni's Platonic love poem and transformed it from an expression on the philosophy of love to one that spoke of the love of philosophers: "Socrates loved out of his chastity not only Alcibiades but almost all of the wisest and handsomest men in Athens . . . all of whom wished to get incitement from the body's outward beauty to look at that of the Soul."[62]

Pico and Benivieni's intellectual coupling signified a precursor to eternal love. Some years later, Benivieni's decision to be buried with Pico symbolized the love and intimacy the two shared. The words Benivieni engraved on Pico's tomb in 1530 read "Here lies Giovanni Pico Mirandola . . . He died in 1494, and lived for 32 years. Girolamo Benivieni, to prevent separate places from disjoining after death the bones of those whose souls were joined by Love while living, provided for this grave." Someone later added some Italian verses: "I pray God Girolamo that in peace also in Heaven you may be joined with your Pico as you were on earth, and as now your dead corpse together with his sacred bones here lies."[63]

The sexual preferences of the Neoplatonic Florentine circle, which counted Pico among its leading members, also had a connection with astrology. Astrologers believed that planetary positions in nativities determined sexual disposition. Leone Ebreo, who spent some time in Florence in the early 1490s, writes in his

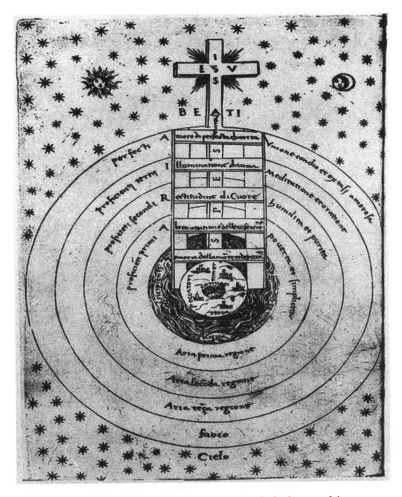

The Ladder of Spiritual Life, correlating the rungs with the letters of the name Maria. Many artists of the time, including Michelangelo in *Madonna of the Steps,* likened Mary to a stairway by which God came down to earth as Jesus and by which mortals in turn may ascend to heaven. The progression from the profane to the divine is represented in this book illustration made by Francesco Rosselli (b. 1448, d. after 1508). Note that the stars are not circumscribed by a sphere but scattered around, indicating the appropriation of the Kabbalist argument about the infinity of the universe. Domenico Benivieni, *Scala della vita spirituale sopra il nome di Maria* (Florence, 1495). Courtesy of Houghton Library, Harvard University.

famous *Dialoghi d'amore* that if the number of masculine signs equals the number of feminine signs in the horoscope, and Venus is the ruling planet, the male child will be a hermaphrodite, which in such context more likely means he will love men and not be shamed to be both active and passive in sexual intercourse.[64] Ebreo's reading notwithstanding, most early modern astrological commentaries were hostile toward sodomy and expressed contempt for male sexual passivity, though the active partner was usually tolerated.

Noted astrologers like Girolamo Manfredi (1430–93), against whom Pico wrote in the *Disputationes*, expressed views opposing sodomy. Pico attacked Manfredi for personal reasons: he had provided false astrological predictions regarding Pico's siblings.[65] But he also took issue with Manfredi's infamous antihomosexual accounts, published in his popular *Libre de homine* (1474, often reprinted as *Il libro del perche*), where he wrote that sodomy and homosexual practices caused infirmity and harm the joints.[66]

Thus astrological accounts saw sodomy, effeminacy, and what were seen as lewd and filthy sexual behavior as *contra naturam*, all resulting from the correlative combinations of planetary or astral influences under which a person was born.[67] Pico did not accept such sexual determinism and used the *Disputationes* to scorn the astrologers' claim that they could predict the gender, the looks, and the sexual inclination of a child.[68] Such astrological predictions targeted not Florence in general but Pico and his friends in particular.

Pico's Loves

Love possessed Pico. For his generation, love represented the force of attraction that held together the universe, nature, and society. After commenting on the biblical Song of Songs with Alemanno, Pico wove Kabbalistic readings of love into the generally Neoplatonic discourse and in 1491 published a short tract, *De ente et uno* (On Being and the One) in which he argues that Being and the One are not separated but are mutually conditioned. The relationship between God and man relies on reciprocal crowning—humans crown God the king of the heavens, and God crowns man the superior creature in nature. Even knowing nature represents a process of coupling with God, a process that climaxes in ultimate happiness and, as in Narbonni's view, ends with the kiss of death (*morte osculi*).[69] Pico gives a poetic description in "The Soul's Search for Peace and Its Union with God," in the *Oration*, and projects it anew onto sociointellectual relations to transform it into a Pythagorean *amicitia* (friendship): the most holy peace is "the indissoluble bond, the harmonious friendship in which all souls, in one mind, a Mind that is above all minds, are not only in agreement but indeed, in a certain ineffable way, inwardly become one." The Pythagoreans posited this friendship, Pico stresses, as the purpose of all philosophy: "Angels who came down to earth announced to men of goodwill so that these men who would practice love to man and to God would ascend to heaven and be transformed by it into angels."[70] Thus for Pico, love and friendship signal types of coupling between human beings as well as a general force of attraction between physical bodies. Because of its natural faculty for love and friendship, the soul progresses toward its union with God.

Later, in his commentary, Pico defines love as desire based on knowledge. "To know things," he states, "is to desire and possess them."[71] To know and to love mean to posses the object; to know nature means to experience it and possess it by embodied interaction, by *naturalis magia,* or practical natural philosophy, as he called it. Indeed, Pico lived according to the dictates of his belief in the exaltation of human friendship as the locus of philosophical achievement. His exchanges and partnerships with close friends represented the highest level of intellectual activity, occurring first with Benivieni, with whom he began an intellectual coupling when he wrote a commentary on his love song. When, after Pico's death, Benivieni published the poem and commentary together, he made implicit reference to their shared forbidden love and forbidden interpretation of love. In his introduction, Benivieni recounts the history of the project, explaining that they decided to suspend publication because "in looking at this poem and commentary again later on, . . . lacking some of that spirit and fervor which led me to compose it and him to interpret it, there was born in our minds a doubt as to whether it was fitting to discuss our Platonic love."[72]

Later Pico and Poliziano shared an erotic-Platonic love, beginning with Pico's defense of Poliziano's Aristotelianism in the face of Lorenzo's and Ficino's Platonic critique and deepening in the early 1490s, when they spent time working closely and making several book-hunting trips. They would travel together for some weeks, moving from one private library to the other, where they browsed pages, discussed their value, and finally collected manuscripts to enrich Lorenzo's library.[73] During one of these trips, in June 1491, they traveled to Bologna, where they met influential people and visited public and private libraries. They bought and copied many books and manuscripts and enjoyed the time together. As Poliziano wrote to a friend, "We slept in the same room together."[74] During the trip, Poliziano received a letter from Niccolò Leoniceno, Pico's teacher from Ferrara, who wrote, "I hear that you pass day and night with our dear Giovanni Pico della Mirandola, in every exercise of the fields of learning, and since, besides, the support and favor of Lorenzo . . . who procures for you every resource for philosophizing freely."⁷⁵ They projected their love onto an exploration of books and knowledge. After Lorenzo's death, the men's relationship further strengthened. While Pico closed himself off in his villa, Poliziano's complained of his absence, expressing his longing for his beloved friend.

At one point, Pico dabbled in poetic writing, titling the resulting love poems, *Amores mei* (My Loves). He sent them to Poliziano, who returned them with some notes referring to their shared love of young boys: "My, you are a wit to match me with your *Loves* and to require that such attractive boys be received by me." Soon thereafter, Pico burned the set of poems, about which Poliziano wrote, "I hear that you have burned the erotic epigrams you wrote some time ago, worried, perhaps, that they might prove detrimental either to your reputation or to others' morality." To recover the burned poems, Poliziano decided to amuse himself by "describing those very *Loves* being consigned to the flames by you" in Greek verse: "Pico, often pierced and set aflame by Loves / More would not bear, but took their every arm, / Bows and arrows, quivers too, and all amassed. / The heap he torched with stolen flames. / They themselves be seized, and tied their tender hands / With strings, and hurled them on the pyre. / Fire he burned with fire: why then, of foolish Loves to Pico, Prince of Muses, did you fly?"⁷⁶

Here, the open-minded poet mocks his philosopher friend for the secret loves of which he is embarrassed to speak. Whereas Poliziano was infamous for his love of boys and explicitly mentioned his admiration of adolescents' bodies, Pico displayed a more concerned and cautious attitude in his *Loves*.

During the same period, Pico also worked on his manuscript-translation of Ḥayy. More than just a philosophical framework that suited his arguments against astrology, he found in the philosophical novel a template for his favored social and intellectual relationships. *Ḥayy Ibn-Yaqẓān* presents a manly utopia about a lonely boy, in which females and sexuality do not appear. When, at the end of the story, Salman and Absal find Ḥayy and bring him back to their socialized and civilized world, Ḥayy quickly learns the language, interacts with their circle of philosophers, and exhibits his intellectual superiority, but is left unchallenged and bored. Ḥayy eventually decides to return to the life of solitude on the island, this time with Absal in tow. The story ends with Ḥayy and Absal on their island, where they have established a semiscientific society of two and live and love happily ever after. The last paragraph of Pico's manuscript-translation reads, "Then it is easy for us to discover the secret Ḥayy had and the way he unveiled the curtain," showing "the corrupt views" of those who "have declared themselves to be philosophers . . . and who have rejected the Kabbalah." Here we find Pico's first mention of the vision Ḥayy observed in his ecstasy as a Kabbalistic secret. Being hidden, Kabbalist secrets can only be reaching through an "increase of [philosophers'] zeal" in pursuit of the ultimate truth and an increase in "their own love." As a consequence, through the love of the truth and of each other, they can "shine the most secret of secrets."[77]

Pico found features in Ḥayy's story that fit his philosophical and cultural preferences. A child prodigy and autodidact, like Pico, Ḥayy also practiced the Platonic love of wisdom and its projection onto masculine relationships between colleagues who explore the truth together. Ḥayy also acquires knowledge about nature by actively exploring passive or dead nature, which allows him to explore, conquer, possess, penetrate, and take control over it. He reaches the climax of knowledge through communion with God and in experiencing that ultimate joy epitomizes the secrets of Kabbalah. Finally, he engages in a Platonic love of the truth, a love eventually projected back onto society, in the form of the erotic love between two colleagues. Moreover, the question of sexuality also had philosophical ramifications, especially concerning Pico's view of the exploration of nature as a coupling of man with nature.

Dead Nature and A Lively Boy

Even before Lorenzo's death, the humanist circle over which he presided divided into two philosophical pillars. Pico viewed Ficino as a friend and an elder scholar but not as a binding intellectual authority. In his commentary on the love poem of Girolamo Benivieni, Pico ridiculed Ficino's argument claiming a distinction between the beautiful and the good, the extrinsic and the intrinsic, by holding that the distinction lies more along the lines of that between the species and its genus.[78] Later, Poliziano wrote a piece on Aristotle's *Nicomachean Ethics* to exemplify his rejection of the Platonist separation of form and matter. As a response to Lorenzo's critique of Poliziano, Pico wrote another short tract, *De ente et uno* (On Being and the One), in 1491. Addressed to Poliziano, whom he called an "almost inseparable companion,"[79] Pico argues that the One is not above Being, as Ficino and other Platonists would say, but equivalent to it. In the positions he espoused during both of these controversies, Pico defended his lovers—first Benivieni and then Poliziano.

Finally, after Lorenzo died in 1492, another controversy sprang up, along with the writing of the *Disputationes* and the translation of *Ḥayy Ibn-Yaqẓān*. At stake was a fundamental question about the nature of the cosmos—alive or dead. It had ramifications for the question of the relation of man to cosmos. If the cosmos is alive and ensouled, discovering it entails a contemplative approach. If the cosmos is dead, the only way to understand nature is through active exploration.

In his canonical work *Platonic Theology* (*Theologia Platonica de immortalitate animorum*), first published in 1482, with a second edition in 1491, Ficino argues that the acquisition of knowledge about a vital nature (*natura vitalis*) goes through "the soul's two principal offices: to contemplate and to deliberate. In both it opposes the bodies."[80] In only one place (book 6, chapter I) does he cite Pico's qualifications on his theory of the soul, arguing that the soul is not an incorporeal light but is dependent on the existence of the body. In respond, Ficino stresses that "fantasy is the best friend of the senses," deploying a sequence of cases in which children acquired firsthand knowledge through their senses, resulting in a deluded view of nature. "Just as a child who is born blind," Ficino continued, "finds it difficult to believe in the existence of a variety of colors and of light, likewise a rational soul surrounded until now by body's darkness can scarcely be brought to accept that an incorporeal light exits."[81]

Ficino further promoted his view of an ensouled nature connected to man through incorporeal light in his book on psychological, pharmacological, and astrological therapy, *De vita libri tres,* published in 1489. He stresses that nature lives by means of the soul of the world, the *anima mundi*. The always-present "incorporeal light" links the tangible part of the body of the world to its soul, creating "a concord of the world and the nature of man under the stars." Thus for Ficino, the "alive cosmos" requires a spiritual coupling with nature. In prescribing the way to receive the supercelestial flow of intellect, he notes that "our spirits can absorb the most possible spirits and the life of the universe, and what planets recreate the spirit and what really belong to each planet."[82] Knowing nature, therefore, required a passive act, accomplished mostly through the *vita contemplativa*, through which man spiritually couples with the active spirit of the cosmos.

Pico, disagreeing, turned away from the macrocosm to the microcosm. In place of Ficino's spiritual coupling of ensouled nature he offered a practical coupling with passive nature. In place of "incorporeal light" Pico placed the physical light at the center of his natural philosophy. In the *Disputationes*, he reduces the effects of the heavenly bodies to mere physical properties, thereby detaching man from predetermined astrological causes. Light, and its property, heat, became the central agents of nature, determining the variations within human beings generated by moisture, hot, dry, and cold qualities created by the angle of the solar ecliptic at different locations. "In an inquiry concerning nativities and individual temperaments in general," Pico writes, "one can see that there are circumstances of no small importance and of no trifling character, which join to cause the special qualities of those who are born." He then gives a description of the conditions for the particular character of the fetus that range from genetics (seeds) to environment (climate and customs):

> For differences of seed [with their specific forms] exert a very great influence on the special traits of the genus, since if the ambient [the celestial configuration] and the horizon [the place with its unique horizon] are the same, each seed prevails to express in general its own form; all these things . . . come to be from vapors which heat alone elicits, therefore, heat makes everything, namely everything universally, since their differentiation and variety do not come to be from the variety of the celestial configuration, as the astrologers fantasize, but with the matter and place in which they are generated."[83]

Later, in book 3 of the *Disputationes,* Pico argues that the planets and stars differ from the sun and moon and do not have any influence or only a weak influence on us. "Influence of Mars, Mercury," and other planets "is less than of the sun and the moon since the greater influence comes from the physical property of light." He pointed out that the fetus gets its main characteristics from human seeds and different climates and not from the influence of the moon or other planets. The influence of temperature creates diverse populations that vary according to their location in southern or more northern latitudes.[84]

Pico found a corroborating view in his manuscript-translation of *Ḥayy Ibn-Yaqẓān,* which describes the island of Wāqwāq as an exceptional place with balanced climatic conditions. "Abubachar said: Our wise ancestors indicated that among the Indian islands there is an island that is situated under the equatorial line. This is the island where man was born without father nor mother, since it is of a more temperate climate than all other places on earth and is more perfect than them because of the shining of the light upon them."[85]

Pico then describes the process of Ḥayy's spontaneous generation, emphasizing the effects of the sun's rays. "Abubachar said: he was born without mother and father, it is said that the belly of the earth of that island had fermented mud within it during the length of a year, and that mud combined wet with dry and warm with cold and its particles interacted equally one on the other." Pico then metaphorically describes the process of generation (echoing the metaphor used in Alemanno's commentary on *Ḥayy*): "And that mud generated a bubble similar to the ones on pots, and the bubbling up was made through the force of mixture. And in the midst of it was produced one very small bubble which is divided into two parts distant from each other, the one is filled with a fine mass made of air at the ultimate limit of temperature which is suitable to it, and following this the spirit which is so blessed."[86]

In bringing up spontaneous generation, Pico proposed a theory of human ancestry that did not incorporate racial variety as the product of one human mother and could reduce the effects of the celestial bodies to the property of heat as the generator of life. Doing so reduced the power of any prediction to a practical exploration of nature and meteorological observations only. To prove his point, Pico compared the weather in his "suburban villa ... in which [could be seen] every major transformation of the air on individual days, with the decrees of the astrologers meanwhile placed before my eyes. The heavens favored me ... In a continual observation of over 130 days, I saw no more than six or seven days such as I have foreseen beforehand in the books they have written."[87]

Pico asserted that human seeds, sun's light, physical heat, and various climates were the main agencies for various human characters. In combination, these factors were responsible for various civilizations, cultures, and customs affecting human character. For instance, in book 3 he talks about the ways climate influences customs, laws, and religion and gives the example of "the northern Gallic people who were known to have a particular sexual interest in young boys."[88]

Pico's arguments against astrology sought to replace the importance of celestial configuration in relation to nativity with an emphasis on the angles of the sun and the range of heat produced in any given location. Human temper, variations in physical properties, and cultural inclinations all depend on the location of human seeds on certain latitudes, where the effects of the sun's radiation vary.

Ḥayy's story also echoes in the section of the *Disputationes* that deals with the historicization of astrology, where Pico placed the origins of astrology in the ancient East. Signaling a change in attitude, he expressed contempt for the Eastern cultures he so admired early in his life, embracing instead, according to Anthony Grafton, the Greek view of human liberty at the end of his life.[89] Pico's attitude cannot be viewed as so straightforward, however, because although he generally professes to detest the Eastern parts of world culture, where astrology had been cultivated, he holds up *Ḥayy* as an exception that not only rejects astrological determinism but also represents a reliable account of the physical creation of and variation among species. Rather than viewing *Ḥayy* as emblematic of its own culture, Pico saw in it a radical anti-astrological program that promoted autodidactic learning reliant on the active and free exploration of nature. In Pico's eyes, Ḥayy represents the quintessential example of an individual who molds his own life.

These revolutionary views about astrology and the relations of man and nature represented, however, the by-product of the cultural revolution that Florence experienced in the wake of Lorenzo's death. Various programs for molding, educating, and mobilizing the adolescents who overflowed the streets of the city were at stake.

From Corrupt Boys to Righteous Boys

It seems likely that the consideration of the relations between man and nature represented a projection of the actual erotic love that Pico and his friends practiced and idealized. The Florentine Platonic circle looked like a cultural institution that not only encouraged self-indulgence through philosophy and art but also a offered a cultural role model for boys who practiced homosexuality on the streets of the city. A few clues have emerged suggesting that the murders of Poliziano and Pico were sexually motivated. Poliziano was killed in November 1494. Various contemporary sources talked about "a shameful and miserable death," but none openly spoke about the details. On October 4th, Antonio Calderini wrote to Niccolò Michelozzi that the murderer left signs on Poliziano's body indicating that his death was punishment for his "insatiable greed in life."[90] Fifty years later, the humanist Pierio Valeriano, who admired Poliziano, wrote his *Hieroglyphica*, in which he tried to clear Poliziano's reputation and at the same time implied the prime suspect in the murder by expressing his satisfaction over Savonarola's execution.[91]

Fewer rumors followed Pico's murder. Just before he died he fulminated against his enemies in the *Disputationes* as "ignorant people like that kind who killed Socrates."[92] Some years after his death, his nephew Gianfrancesco, in his *Life of Pico*, simply writes that "[Pico] was suddenly taken with a fever which crept into the interior parts of his body. Neither the medicine nor the doctors sent by King Charles of France, then camped outside of Florence, could recover him and he died after three days."[93] After Pico's murder, Savonarola gave the funeral eulogy. "Oh city of Florence," he loudly harangued, "I have a secret to tell you ... Pico was oft-times conversant with me and revealed to me the secrets of his heart."

Pico's secrets, according to Savonarola, involved his deep worries about eternal damnation for his sins. For two years Savonarola had "threatened" Pico with punishment were he not to follow his calling to serve God and change his ways. "I prayed to God myself that he might be somewhat beaten to compel him to take that way that God from above showed him. But I desired not this scourge upon him that he was beaten with." Savonarola merely hoped that Pico would learn from such beating to stop pursuing the pleasures of this life and start living as a

The self-indulging philosopher and the ascetic Dominican friar. Although Pico, *left*, was surrounded by various artists, this is the only portrait from his lifetime. The anonymous painter presents Pico as a long-haired colorful figure with a somewhat feminine twist. Uffizi, Florence. Girolamo Savonarola's portrait, *right*, was made by the Dominican painter Fra Bartolomeo. The sharp profile of the ascetic, under a dark hood, is painted against a black background. The Latin inscription on the panel below the portrait indicates the prophetic pretension: "Portrait of the prophet Jerome of Ferrara, sent by God." Courtesy of Museo di San Marco, Florence.

righteous Christian. "I pray that he would not be doomed to perpetual suffering in hell" but only "suffer the fire of purgatory for a short time." Concluding his eulogy, Savonarola indicated that the length of time consigned to purgatory was contingent upon the prayers, charities, and good deeds of his friends. "After this period his venial offenses will be cleansed," and Pico will shortly thereafter "enter the inaccessible and infinite light of heaven."[94]

The years following Pico and Poliziano's murders marked a watershed moment in Florentine history, when a radical transformation in moral, political, and educational attitudes occurred. Under Savonarola's rule, Florence came to know the sight of righteous brotherhoods of adolescents enforcing law and order. In less than five years, Sodom was transformed into a holy city.

Pico had convinced Lorenzo to invite Savonarola to Florence in 1491, a move that reinforced their anti-Rome approach by bringing to the city the man who established his reputation defying the ecclesiastic establishment. Despite their shared attitudes toward papal supremacy, Pico and Savonarola soon found their cultural views diverging. Whereas Pico inclined toward the endorsement of the self-indulging life, practiced homosexuality, and maintained a close friendship with a man infamous for sodomizing young boys, Savonarola encouraged asceticism and virtue. Pico's optimistic preoccupation with the creation, the vision of the first man in paradise before the Fall, directly communicating with God and nature, stood in stark contrast to Savonarola's dark and apocalyptic vision of heaven and hell. In his anti-astrological text, Pico emphasized the role of firsthand experience as determinant of human disposition and thought that by use of their free will boys can independently mold their lives. Savonarola, on the other hand, was preoccupied with destruction and the end of the world. He thought that the process of self-molding could lead boys to such perverse phenomena as sodomy. He therefore stressed that boys should be instructed by righteous teachers who would show them the what he saw as the right way of life and would mobilize them to serve the public good. Soon after Pico's murder, when Savonarola took over Florence, cultural practices that emphasized asceticism and predestination prevailed on the streets of the city.

Savonarola aimed to create a perfect political community, based on probity, reason, and the goodwill of the citizens, by whose virtues the political machine would work with no need for tyrants or aristocracy. He laid out the program to achieve this goal in a speech on December 12, 1494, less than a month after Pico's murder. "Man does not remain alone," Savonarola opined, stressing man's social nature. Whereas animals can live alone and provide for their needs by natural instinct, "God has given man reason" so that he can provide for his own needs. Men have joined together in cities or towns to "form a community for the common need each had for the other."

Distinguishing between political forms, Savonarola elaborated upon the correlation between geographical and climatic conditions and political cultures. "In the warm parts of this hemisphere, men are more pusillanimous than in other places because they are less sanguine, and thus, in those places, the people easily let themselves be ruled by a single leader." But in the colder northern parts, "where people are more sanguine and less intellectual, they are likewise steadfast and submissive to their one lord and head. But in the middle part, such as Italy, where both the sanguine and the intellectual abound, men do not remain patiently under a single leader, rather each of them would like to be the leader." In these kinds of places, it would be most appropriate to have "a government by the many" regulated by the individual virtues of the public. "Oh Florence," Savonarola preached, "do first those two things—first, everyone goes to confession . . . to be purified of sins," and second "let everyone attend to the common good of the city."[95]

His first goal was to sweep sodomy off the streets of Florence. He urged the Signory of Florence to pass a law against "that accursed vice of sodomy, for which you know that Florence is infamous throughout the whole of Italy. . . . Pass a law, I say, and let it be without mercy; that is, let these people be stoned and burned."[96]

Next, he undertook to reform the education of the local youth in an attempt to transform their behavior from corrupt to righteously Christian. Part of the discussion concerned their activities on the streets of Florence. The centrality of these young men in Florentine street culture can be seen in descriptions of Carnival, which fell on February 16, 1496, demonstrating the radical shift that occurred in Florence after the moralist Savonarolian revolution had begun.

In *A Florentine Diary*, Luca Landucci writes that thousands of boys forcibly cleaned the streets of Florence from vanities like "gambling tables and many vain things seduced by women, sooner did the gamblers hear that the boys of Savonarola were coming then they fled, nor was there a single woman who dared go out not modestly."[97] Paolo de Somenzi, the ambassador of Milan in Florence, wrote in a letter to Lodovico Sforza, duke of Milan, of a procession of "these boys, who numbered about ten thousand, and most of them have not attained fourteen years of age; of six- to nine-year-olds there were some four thousand. First they had a Mass said in the main church with great solemnity, and then these children, separated by quarters, with trumpets before them and crying out, 'Long Live Christ,' went in a procession."[98] In a contemporary *Vita* of Savonarola, Fra Timoteo Bottonio writes that the transformation of the street culture in Florence had been "considered impossible by everyone, because the children had been nourished in many bad habits for a long time, having broad license and liberty to be enveloped in all the vices. (They looked like girls in all their ornaments and coiffures, or rather like public prostitutes, shameless in word and deed), and especially in that unspeakable vice—a thing abominable to name."[99]

The *Disputationes* alludes to Pico's contradicting views. He could look to Ḥayy to as a model of moderation between extreme licentiousness, on the one hand,

and Savonarola's grim moral rectitude, on the other. Savonarola, however, insisted on editing the already edited papers of Pico's *Disputationes* and, turning away from Pico's emphasis on man's free will as disproving both determinism and astrology, placed the emphasis on the changing will of God. Whereas Pico placed an active man at the center of a dead cosmos, Savonarola used the refutation of astrology for religious purposes, consigning to prophets, like himself, the task of manifesting the will of God.

Although never printed, Pico's manuscript-translation of *Ḥayy Ibn-Yaqẓān* soon had an impact on local Renaissance culture. Antonio Fregoso (d. 1515), a contemporary Italian poet, read Pico's manuscript-translation and included a poem, *De lo istinto natural*, in the 1525 edition of his *Opere*. The poem speaks of an autodidact on a desert island who explores nature through trial and error, dissection, and experiment.[100]

> There is an island somewhere in the wide ocean
> That has not yet been inhabited by human offspring
> For this we believe the Hebrew scholars
> It warms itself up with the wonderful sun
> And there it is almost always spring
> And at all time flowers grow
> There is no ice, no snow and no awful frozen dew
> Nor is it extremely hot to spoil things.
>
> A young boy who lived alone in this beautiful island . . .
> had a natural instinct and capability to learn by himself.
> But though he was well fed on that strange and lonely place
> The ingenuity of his clothes was clear, which demonstrated to them who found him, that
> he was equipped with reason and intellect.

The boy "had been created without human seed" in a process reminiscent of Pico's description of the physical influence of the sun's rays.

> By the bright sun with his temperate rays
> It had prepared itself naturally
> To become the procreator of this human fetus
> She had been "impregnated" by the rays of the sun
> And she swelled like a uterus
> A little belly in her had grown

Which was fed by her roots
And when the time was ripe
The fetus received a shape and life involuntarily
The rays like a flame penetrated the soil
The rays lighten and shine, the soil feeds it
So like the sun from which all life descends
So did the vital life in him descend
The light that makes him hear and understand
And gives him intellect
And when mature at due time
He came to the world like a lighted oil lamp.

Just as Ḥayy dissected his mother-gazelle to find the cause of her death, Fregoso's autodidact dissects a gazelle to explore nature, in the process discovering the principles of anatomy:

Eager to learn what was inside
He opened her while in tears
To find the cause of that vital noise
And where he heard repeatedly that natural beating when she was alive
There he started to investigate
He first opened the left side and
Found the heart and when looking further
That which from the heart spreads:
Seeing her cells and arteries
Made with such a wonderful craftsmanship
He was astonished by that work of art
So by himself he understood when leaving the room [metaphor for the body]
Who was the master of the building [metaphor for the body]
He understood by way of his sight and ears
By way of his smell and other instruments
That the heart is the master of the whole that obeys
That all the nerves veins, and instruments
Of the mortal body are moved by it
Seeing then the machinery of the bones
Connected with so much art by the wise and wonderful architect
Then his thoughts went in another direction
And he imagined that the light that time distributed
Sent her that interior vital heat. . . .

Fregoso here envisions dissection as a self-directed process that uses what Pico would have called practical natural philosophy, which relies solely on natural instincts: "By this evidence, expressed by rational thinking / . . . without needing someone to teach him [*ingegno*] / . . . he was full of natural science."[101] Thus Fregoso, following Pico and *Ḥayy*, establishes man's dignity and superiority in his ability to rely on reason, to use his natural instincts, and to explore and conquer passive nature.

Pico's influence spread beyond the Italian peninsula. Thomas More became his most influential follower; he was conversant with Pico's oeuvre, and in 1510 he translated Pico's *Vita* (written by his nephew Gianfrancesco) from Latin to English. More's *Utopia* seems to echo some of Pico's fascination with a state of the Adamic man, subject only to nature and God, with no social or political constraints. Other utopian writers, such as Tommaso Campanella and Francis Bacon, both of whom depicted utopian island communities where men practice *naturalis magia*, also found in Pico a source of inspiration. In bringing such utopian visions into practice, Federico Cesi, founder of the Lincei Society, reflected on the experience of the Florentine Platonic Academy that privileged male chaste love over heterosexual desire. As Mario Biagioli shows, the *Lynceographum*, the academy's proposed statutes and bylaws (1614), required the avoidance of "the attractions of Venus," "bad women and profane love," "Venereal lust," "prostitutes," "tempting lust," "low passions of the body," "carnal drives," "libidinous excitements," and "the body's inane desires." Additionally, the Lynceographum ordered the academicians to steer clear of "scandals with boys" and legislated how violations of the Lincei's code of honor were to be reprimanded and punished.[102]

These utopian visions later served as the core idea in the rise of experimentalism. Francis Bacon, in *Instauration magna* (*The Great Instauration*, 1620), introduced his seventeenth-century readers to a systematic application of self-learning on the exploration of nature. Bacon discouraged his readers from forming an "opinion or judgment either out of the crowd authorities, or out of the forms of demonstrations (which have now acquired a sanction like that of judicial laws)." Rather, he instructed that the seeker of knowledge ought to create trials to "familiarize his thoughts with that subtlety of nature to which experience bears witness. Let him correct by seasonable patience and due delay and depraved and deep-rooted habit of his mind. And when all this is done and he has begun to be his own master, let him (if he will) use his own judgment."[103]

Bacon extended an invitation to those who are not "content to rest in the use of knowledge which has already been discovered, and are aspired to penetrate further; to overcome, not an adversary in argument, but nature in action; to seek, not pretty and probable conjectures, but certain and demonstrable knowledge" to join him in passing through "the outer courts of nature" to find a way into "her inner chambers."[104] He called on seventeenth-century English natural philosophers to carry out the application of experimentalism on the various sciences. Such followers later echoed Pico's rewording of Horace's autodidactic statement, *Nullius addictus judicare in verba magestri,* "Not compelled to swear to any master's words," and turned it into the Royal Society's motto, *Nullius in verba.* Society members promoted experimentalism by taking it as their pledge to rely on no book, accept no written authority, but trust only firsthand practical experience of nature.

CHAPTER FOUR

Employing the Self and Experimenting with Nature

Oxford, 1671

ARABIC AND HEBREW MANUSCRIPTS of *Ḥayy Ibn-Yaqẓān* circulated across the Mediterranean throughout the seventeenth century. During the 1630s Edward Pococke traveled twice to the Near East. While in Aleppo (Syria) he came across an Arabic manuscript of *Ḥayy Ibn-Yaqẓān*. Now kept in the Pococke collection at the Bodleian Library, the manuscript has a Hebrew title written in the margin of its title page and is full of Pococke's Latin marginalia. Pococke brought the manuscript back to Oxford in the early 1640s, where it lay in his library for more than three decades until certain circumstances prompted him in 1671 to turn out a Latin translation—*Philosophus autodidactus*, with Latin and Arabic texts on facing pages and with a long introduction. The treatise, however, did not come out in a cultural void. After two centuries of utopian writings, continuous debates over the question of inductive reasoning, and a recent increase in experimentalist writings, the publication corresponded with prevailing approaches to the study of nature. What may seem like an anachronistic publication, however, lay at the heart of the controversy between royalist philosophers like Thomas Hobbes, who deduced arguments from first principles, and parliamentarians like Robert Boyle, who stressed the self-reliance of experimentalists, using autodidacticism as a chief component for empiricist philosophy.

In August 1630, Edward Pococke, a short and gloomy young man, embarked on the *Cheruvim*, one of the Levant Company's ships headed to the Near East. By October the ship had arrived at the port of Beirut; from there Pococke joined a caravan of merchants heading to the Levant Company's factory in Aleppo, where he served as chaplain for five years. One summer afternoon in 1634 he observed and researched the life of the chameleon, which, as he noted, lives in solitude and takes control of its destiny by assimilating into the changing environment.

Around the same time, local colleagues introduced him to *Ḥayy Ibn-Yaqẓān*. Ḥayy's story, with its motifs of solitude and autodidacticism, captivated Pococke and seemed to replicate his own living conditions. Forty years later, at the end of his career in Oxford, he believed that the story would also fascinate readers in England.

Some insights into the character of these potential readers are embedded in the very title of Pococke's work and in its note to the reader. In place of the original title, *Ḥayy Ibn-Yaqẓān*, Pococke's phraseology revealed something both about himself and about what would attract readers to the book. The full title, as it has come down to us, is *Philosophus autodidactus, sive, Epistola Abi Jaafar ebn Tophail de Hai ebn Yokdhan microform: In quâ ostenditur quomodo ex inferiorum contemplatione ad superiorum notitiam ratio humana ascendere possit / ex Arabicâ in linguam Latinam versa ab Edvardo Pocockio* . . . (The autodidactic philosopher, or Epistle of Abū Ja'afar Ibn-Tufayl concerning *Ḥayy Ibn-Yaqẓān*, whereby to show how from contemplation from below human reason ascends to higher ideas).

The notion of the autodidact dated to late antiquity, especially to the writings of Philo; but in the early modern time, Thomas More was the first to use it in England. Against the backdrop of the increasing call among Protestants for a *sola scriptura*, sole reliance on the scriptures, More's 1534 *De tristitia Christi* complained that "there are springing up from day to day, almost like swarms of wasps or hornets, people who boast that they are 'autodidacts' (to use St. Jerome's word) and that, without the commentaries of the old doctors, they find clear, open, and easy all those things which all the ancient fathers confessed they found quite difficult."[1] Although More was mostly concerned about Protestants who disregarded tradition and mediation in their reading of the Bible, he later appropriated autodidactic motifs to his *Utopia*.

But the autodidactic life went beyond theological questions. Before and during More's day, arguments for the nonmediated acquisition of knowledge came from both popular culture and mysticism. Later, in the sixteenth and seventeenth centuries, autodidacts were concerned with reading the book of nature directly, which led to arguments for experimental philosophy. The Royal Society of London for the Improvement of Natural Knowledge (established in 1660) took Horace's *Nullius in verba* as its motto and committed itself to trust "nobody's word," withstanding the domination of authority and calling to verify all statements by an appeal to facts as determined by firsthand experience and experiment.

These experimentalist circles received *Philosophus autodidactus* quite well

Title page, Edward Pococke's *Philosophus autodidactus* (1671). Courtesy of the Clark Library, Los Angeles.

when they came to know of it a few months after its publication in July 1671. The review in *Philosophical Transactions of the Royal Society* introduced the piece and implicitly promoted it to potential experimentalists readers:

> The design of the book is to show, how from contemplation of the things here below, man by the right use of his reason may raise himself into the knowledge of higher things; which is here performed by a feigned history of an infant . . . on an Island not inhabited; where he was nursed up by a gazelle, and coming

First page in bilingual *Philosophus autodidactus* (Oxford, 1671). Courtesy of the Clark Library, Los Angeles.

afterwards to years of knowledge, did by his single use of reason and experience and without any human converse attained the understanding, first of Common things, the necessaries of human life . . . then to the knowledge of Natural things, of Moral, of Divine, etc.[2]

Pococke recast Ḥayy Ibn-Yaqẓān as an oracle from the past, confirming experimentalism and empiricism. He deliberately directed his effort toward this kind of reception and used *Philosophus autodidactus* to communicate with his desired audience—readers already exposed to arguments in favor of inductive reasoning in the exploration of nature—placing particular emphasis on the methodological value of the work as a self-conscious aid to contemporary practices of science, especially the new philosophical venture of experimentalism. The note to the reader, which, though written in Latin, contains phrases in Arabic, Hebrew, and Greek, states that Ḥayy "will exhibit the ultimate strength of reason in the affairs of a nature so inferior, not by scrutinizing and investigating those things which are beyond nature, but when he has employed himself as his own-teacher."[3]

The particular goal of the work is to prescribe "a method for those given over to this sort of contemplation: and, indeed, this has stood out in that method, which after a number of centuries can perhaps still seem new, and the path not tread by others." "In fact," Pocock continues in his note to the reader, Ibn-Tufayl presented a method that unveils natural things that "neither the eye can see, nor the ear hear, nor the human mind perceive."[4] By trial and error Ḥayy was brought to induce the fundamental laws of nature. The gradual acquisition of knowledge that sits in a grand hierarchy—from physics to astronomy and metaphysics—is the prerequisite step that promoted him to the ultimate goal of discovering "the knowledge of God and of the upper world."[5]

Apparently the participation in practices of world trade and traveling enabled Pococke not only to fruitfully converge two rising intellectual trends, local English experimentalism and orientalism.[6] Given his proximate relations with a network of Oxford scholars, Pococke might well have intended his translation of *Ḥayy Ibn-Yaqẓān* to serve as a tool in the hands of contemporary experimentalists and philosophers. The translation highlighted autodidacticism as a central component of experimentalism, facilitating its extension beyond natural philosophy to epistemology, theology, and education. By the turn of the eighteenth century *Philosophus autodidactus* had sparked contagious enthusiasm across Europe, which sent it to print in various languages and in manifold editions, setting it out as the quintessential forerunner of the Enlightenment.

Collecting in the East and Building an Academic Career

Expanding networks of trade facilitated access to Near Eastern sources that eventually affected English intellectual culture. His interest in the Near East was spurred by the growing English effort to pave a way in the Mediterranean for avenues of trade with India and Asia. Between the Crusades and the late sixteenth century, the English, occupied with exploration and gaining hegemony in the North Sea and the Atlantic Ocean, had no access to the Mediterranean and almost no contact with the Near East, and during most of the sixteenth century, Englishmen conducted trade in the Mediterranean through Venetian intermediaries. Queen Elizabeth I, however, aspired to avoid the high costs charged by the Venetians and thus promoted direct connection with the Ottomans. In 1579 she exchanged letters with Sultan Murād III, and in 1580 William Harborne, an English entrepreneur, obtained "capitulations" granting privileges to English merchants working in the Ottoman Empire, similar to those accorded to the French. In formal gratitude, Queen Elizabeth sent Thomas Dallam, a designer of musical instruments, to build an organ for Sultan Murād III, initiating a cross-cultural gift exchange of both art and science with the Ottoman court.[7] A few merchants subsequently formed the Levant Company in London, and a year later Harborne returned to Constantinople as the first English ambassador, bearing gifts, including expensive clocks and other automata.

Local representatives handled trade in Levant's factories in Constantinople and Aleppo. They looked for wine, cotton, and silk from the eastern Mediterranean to exchange for silver that was extracted in the New World. Meanwhile, back home, Levant factories stimulated excitement and interest in Near Eastern cultures. Charles Robson, the first chaplain of the Levant Company in Aleppo, published a pamphlet in 1628 titled *News from Aleppo* in which he enthusiastically wrote that Aleppo is "an Epitome of the whole world . . . English, French, Dutch, Italian, Jews, Greeks, Persians, Moors, Indians, . . . Men of all Countries, of all Religions live there . . . the description of whose different customs in their conversation, and tenets in their Religion, deserves rather a volume than a letter . . . I will only acquaint you with some observations of not ordinary things that are, and have happened in and about *Aleppo*."[8]

In the unique cosmopolitan environment of Aleppo the self-reliant English factory functioned independently as a colonial polity subject to English law, a political arrangement worked out not only through the earlier Ottoman grant of privileges to the English but also through the 1605 trading charter that King James I extended to the Levant Company. According to the charter, Levant was to be considered a "company of merchants of England trading into the Levant Seas and shall be at all times hereafter one fellowship and one body corporate and capable in law."[9] A self-operating polity, it observed its own laws within the boundaries of particular cultural islands: ships in motion on the Mediterranean and factories in Aleppo and Constantinople. Within the factories, representatives of the company conducted everyday practices of trade with locals, and in passing on their cultural experience they eventually affected English culture and politics.[10]

Although Pocock spent significant time in Aleppo, little evidence about that time emerges in Leonard Twells's 1740 biography, which relied on no-longer-extant papers and letters.[11] After finishing his training at Oxford in Near Eastern languages under the guidance of Matthias Pasor, he succeeded Robson as chaplain of the Levant Company in March of 1630, carrying out mundane ministerial duties in Aleppo and, guided by local teachers, continuing his linguistic studies

Employing the Self and Experimenting with Nature 107

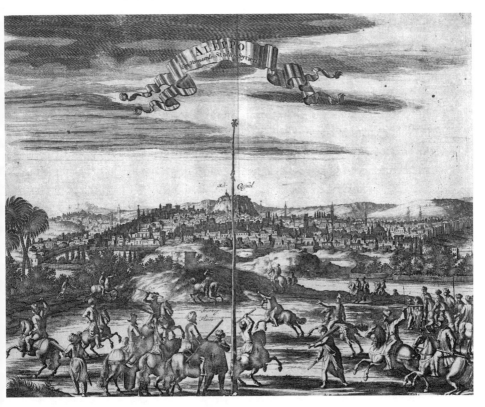

Colorful event in cosmopolitan Aleppo of the 1670s, as illustrated by Olfert Dapper, a Dutch historian and traveler, in his book, *Naukeurige beschryving van gantsch Syrie, en Palestyn of Heilige Lant* (Amsterdam, 1677). Courtesy of Houghton Library, Harvard University.

of Arabic, Hebrew, Syriac, and Ethiopic. Manuscript collecting excited him. Writings of the primitive church, as well as for Syriac versions of the Bible, Hebrew commentaries on the Bible, and geographical and historical works, stretched his motives beyond mere taste and intellectual inclination and on to the political and theological agendas in England on the eve of the Civil War.

In a letter dated October 1631, the chancellor of Oxford, William Laud—later the archbishop of Canterbury and Pococke's patron during the late 1630s—instructed him to purchase manuscripts of the primitive church in Greek and oriental languages.[12] Following Laud's order requiring that every returning ship of the Levant Company deliver one Arabic or Persian manuscript, Pococke supervised the dispatch of manuscripts he and others collected in Aleppo. Other intellectual projects employed Pococke as a collector. Brian Walton, who was putting together the London polyglot Bible, engaged him in the grand project, for which he required various versions of the Bible; Archbishop Ussher employed him for his biblical history projects, which required geographical and historical works in Arabic that would help determine the locations of ancient biblical and mythical sites. Manuscript collecting, however, also played a part in a larger, international race to obtain rare manuscripts exchanged, purchased, and copied by various European representatives—Dutch, Venetian, and French.[13]

Collecting and studying entailed social connections with locals that continued even after he returned to England. His student and successor in Aleppo, Robert Huntington, who gathered botanical and zoological specimens for Boyle and Locke,[14] wrote in 1670 to Pococke about the death of his Arabic teacher, Shaykh Fatahallah, who spoke Pococke's name "with his last breath."[15] Huntington also delivered letters sent from a local agent, Darwish Ahmad, who purchased and copied manuscripts for him. "Since you left us," Darwish writes, "we have been as though our own brother had left us, or the spirit which is in the heart. And therefore, we had sincere joy when we heard the news of your health, and your arrival in your country." He later describes the manuscripts he had transcribed or purchased for him and requested "a little token of what your country affords" in exchange. His letter ends with a request: "Please, send us also printed books of geography."[16]

Activities in Aleppo also combined the contemplative religious life of a chaplain with the exploration of local natural phenomena. In 1634, when a plague afflicted Aleppo and most of the European residents escaped to the countryside, Pococke stayed alone in town and put his "confidence in the Providence of God, ready to meet his good pleasure." He further expressed his solitary life in letters to his friend Thomas Greaves, contrasting his social life at home with seclusion in exile and stating that "there is nothing that may make man envy a traveler."[17]

In addition to his chaplaincy duties, Pococke diverted himself with inquiring into matters of natural history, as well as into local phenomena that were not to be found in his own country. He followed Pico's famous question in the *Oratio*, "Who would not wonder at this chameleon?"—aiming to show that man can alter his nature as the chameleon changes its colors. In a letter to Greaves, he described the wonders of the chameleon, noted its changing colors, and corrected the misconception put forth by Pliny and some other ancient natural historians in claiming that it lives wholly upon air. To come to this conclusion, he conducted a small experiment in which he kept the chameleon in a box for a long time and discovered that it could indeed live for several months without any food but that it would eventually die.[18]

Local phenomena intrigued Europeans because the locales of the Bible dotted the Near East. Pococke adopted a program of reading certain difficulties of scripture against the backdrop of nature as he encountered it in these new and storied surroundings. For instance, he was fascinated by biblical creatures, such as the *tannim*, which most translators rendered as "dragons." Curious about natural historians' opinions that dragons possessed only the vocal power of hissing, he contrasted that with the biblical rendering of their howling and wailing.[19] Through localized readings of biblical passages, those creatures were revealed as wild dogs—jackals, and not dragons. The incorporation of firsthand observations into his commentary on the book of Micah offered a new approach to clarify potential biblical inconsistencies—a combination of natural observation and hermeneutic interpretation.[20]

Such activities were cut short when political theology employed orientalism and experts like Pococke were needed to translate works of the primitive church, legitimating the rising Anglican theology. Laud was convinced that oriental studies in England should be expanded, so as to recover the writings of the primitive church and its latent historical connection, through the Greek Orthodox, to the Anglican church. He established a professorship of Arabic at Oxford and ordered Pococke to return home to take the Oxford chair.[21]

In August Pococke delivered his inaugural lecture on the Arabs' love of poetry. Laud's statute required the professor to lecture on Arabic grammar and literature every Wednesday to all bachelors and students in medicine. But Pococke stayed in Oxford for only one more year before he left again for the Near East.

Although it may seem odd that he would return to a place of such personal hardship, academic politics and political theology combined to make the move attractive to him. In December of 1637 John Greaves, a fellow at Merton College and later the Salivian Chair of Astronomy at Oxford (and the brother of Thomas Greaves), returned to England from Italy and prepared to leave again for the Near East to conduct observations from the locales of ancient astronomers. He asked Pococke to join him as consul. In returning to the East and procuring treatises useful to English readers Pococke had hoped to build up an academic reputation and to secure his position at Oxford.[22] Whereas he had previously collected what he liked, on this visit he collected what would appeal to others.

The second trip went beyond Aleppo. Under Laud's patronage and instructions, Greaves and Pococke went first to Constantinople and collected manuscripts of the Greek Orthodox Church. Soon thereafter, they parted company, Greaves going on to Rhodes and Alexandria to corroborate the latitudes of the Ptolemaic tables while Pococke stayed in Constantinople for the next three years, living in Galata in the house of the English ambassador, Peter Wyche.

To further improve his linguistic knowledge and collect manuscripts, Pococke turned to several learned Jews, notably Jacob Romano, a Sephardic scholar who introduced Pococke to the study of Judeo-Arabic and to the late medieval writings of Mordechai Comtiano and Moses Narbonni, both of whom, in an interesting convergence, had penned Hebrew commentaries on Ibn-Tufayl's *Hayy Ibn-Yaqzān*. This connection may explain the Hebrew title on Pococke's manuscript, indicating that Pococke was searching for other versions of the Andalusian treatise.

After spending three years in the Near East, Pococke headed back home and made a short stop in Paris to meet Hugo Grotius, expressing his admiration for his works. On his return to Oxford he got caught up in the politics of the Civil War, visiting his patron Laud, then imprisoned in the Tower of London. Inspired by the example of Hugo Grotius, who, assisted by his wife, had escaped Calvinist jail, Pococke tried, without avail, to convince Laud to escape. Instead, Laud was beheaded, and the political stance of the authoritarian academic-royal milieu further worsened. With the execution of King Charles I, John Greaves, his friend and royalist ally, wrote to Pococke in a tone of utter despair, "O! my good friend, my good friend! Never was there sorrow like our sorrow! What a perpetual infamy will stick on our religion and nation! . . .—O Lord God, if it be thy blessed will, have mercy upon us not according to our merits, but thy mercies, and remove this great sin, and thy Judgments, from the nation."[23] In the aftermath of the king's execution, royalists such as Greaves were ejected from Oxford while parliamentarians filled the university and started pushing an experimentalist program.

In December 1650, Pococke's own position was in danger. The purges of royalist scholars by the committee of Parliament in charge of regulating the universities affected scholarship at Oxford. The committee acknowledged the necessity of having a prominent scholar in Hebrew and Arabic and searched for a proper successor to Pococke among various Jewish and Christian scholars.[24] Some leading figures felt uneasy at the thought of losing Pococke. Among those concerned were John Wilkins and John Wallis, among others, who in November 1650 signed a petition to the committee of Parliament asking that Pococke be kept in his position for his valuable knowledge and training in Middle Eastern cultures.[25]

Despite his political loyalties, Pococke remained the only royalist left among the scholars at Oxford. These particular circumstances formed an even stronger commitment to remain isolated, to view events "from below" the hegemonic politics of academia.

Academic survival required a publishing strategy. Works that resonated with the interests of his local audience were prioritized. He made use of the fruits of his collection and prepared to print some of the numerous manuscripts he had brought back from Aleppo and Constantinople. His early publications do not indicate a specific philosophical choice and served the interest of a wide audience. Works such as *History of the Dynasties of Abu'l-Faraj* (1650) and *Porta Mosis* (1655), a bilingual edition of Judeo-Arabic and Latin translations of Maimonides' commentary on the Mishna.

To fortify his academic position, Pococke looked for ways to approach the foremost experimentalist, Robert Boyle, and to interest him in his works. He developed a good sense of Boyle's interests and even personal inclinations, eventually falling into a pattern of first beginning a translation and only later trying to gain Boyle's patronage of it by appealing to his religious, philosophical, and cultural interests. From Boyle's vantage, Pococke seemed a potential link to the mysterious cultures of the Near East. The collaboration and friendship between the two started over coffee, the newest fashion to be imported from the Near East. Boyle sponsored Pococke to translate an Arabic text on the culture of coffee drinking in the Near East, titling the short work by sixteenth-century Arab physician Da'ud ibn 'Umar Antaki, *The nature of the drink kauhi, or coffe*—the first European treatise to discuss the origins of coffee and its effects upon health.[26]

Seeking Consensus and Trustworthy Testimonies

The fresh partnership soon moved from cultural to political issues, from coffee to reliable testimonies. During the 1660s and 1670s, Boyle employed Pococke to circulate conversion books and to procure and print manuscripts from the East with exotic appeal to the general public.[27] Sponsored by Robert Boyle, Pococke published in 1660 an Arabic translation of Hugo Grotius's famous polemic, *De veritate religionis Christianae*, seeing it as a good means for converting Muslims.

In his rich study on Christian apologias in early modern times, François Laplanche has listed Grotius's *De veritate religionis Christianae* in a category of writings that opposed any division between faith and reason and stressed the importance of a natural and rational knowledge of God.[28] Grotius, an officer of the Dutch Indian Company and a utopian political thinker, proposed that international law stand as the political frame of reference for obtaining peace on the seas. He had to suffer the impact of dogma when Calvinists imprisoned him in 1618 for issuing an edict asserting that only the basic tenets necessary for undergirding civil order—the existence of God and his providence—ought to be enforced, while differences on obscure theological doctrines should be left to private conscience. Unlike authoritative superimposed Catholic dogma or the independent Protestant reading of the Bible, Grotius frequently advocated religious authority that derived from consensus.

In the first book of *De veritate*, Grotius based his proof of God's existence on the obvious agreement of all people (*consensus gentium*) that God represents the "first cause." He made the political order equivalent to the natural order and argued that the maintenance of that order, in spite of changes in government and their forms, represented proof of God's providence. In the second book of *De veritate*, he stated that though differences of opinion exist in all arts and sciences, they always remain within fixed bounds, because usually some sort of agreement exists on the central points of inquiry. To achieve agreements, those involved need to acknowledge that various kinds of proofs can be achieved and that the proofs required in mathematics are different from proofs concerning fact, since the latter necessitates trustworthy testimonies and consensus.[29]

In their meeting in Paris, Pococke expressed an appreciation of Grotius's work and its potential for propagating Christianity in the Near East. Grotius reported on the meeting in a letter to his brother, saying that Pococke was interested in translating his apologia into Arabic because his experience in the East convinced him that no better book to gain converts among Muslims could be found.[30] At the same time, the clash between England's Parliament and its king was heading toward a climax. The arguments of Christianity's truth as deriving from unity fitted Pococke's royalist belief that theological and philosophical controversies derived from arguments over faith rather than the practice of ethics. For him the central question remained liturgy, not dogma.[31]

Although Pococke had almost completed the translation, the Civil War was not a good time for publishing it. Once Boyle was brought on as patron in 1660, however, the work could be published and distributed for free in the Near East. Pococke delivered a copy of the translation to Grotius and tried to make use of Boyle's prestige. "If you find in it anything to answer expectation," he asked of Boyle, "your good word and approbation will help vindicate it from contempt." When distributed among Muslims, however, the Arabic edition would have to be free of any names or identifying information about its sponsors, "for I think it will be better that any which should be sent into those Eastern parts, should go without any name or title of Persons or men, under that pretence, be hindered. But if you please to order it otherwise, your commands shall be observed." Pococke also advised Boyle about types of Eastern binding they should imitate, in order to promote the idea of the book as a local production.[32]

The project had a far-reaching scope and involved several individuals also in Aleppo. Robert Huntington, chaplain of the Levant Company in Aleppo at the time, wrote to Boyle on September 1673, thanking him for the service "done to our religion by setting forth Grotius in Arabic" and urging Boyle to give Pococke other translation projects.[33] Boyle answered in December, thanking Huntington for his services and confessing,

> I have a great deal of satisfaction to find ground to hope by what you tell me, that my poor Endeavors to promote the Gospel in those parts are not like to prove altogether fruitless, and thus they were first & chiefly designed for the conversion of Infidels ... I have therefore by the Merchant to whom you directed me sent you 3 dozen more of those Arabic books bound as the others were, being not solicitous to exceed that number now, both because I had a short warning given me of the departure of the ships and because of the danger of the seas, but I hope hereafter to send a further supply.[34]

Boyle's interest in *De veritate* went beyond the mere commitment to propagating Christianity. *De veritate* held a special appeal for him, arguing for an antidogmatic theology constituted of logic and consensus among participants. Boyle was particularly drawn to Grotius's explication of miracles. Grotius argued that the truth of Christianity cannot be made plausible merely through natural or rational arguments but must rely on the irrefutable testimonies of Christ's resurrection and the miracles performed by Christ and his apostles. Miracles performed by Jesus were true, Grotius asserted, because they were "founded upon sufficient and credible testimonies." The resurrection of Jesus truly happened because credible witnesses "could not fully persuade men, unless they affirmed themselves to be eye-witnesses of it." Miracles stand or fall on the reliability of the traditional testimonies.[35] Using Grotius as a source of inspiration, Boyle answered critiques of experimentalism and stressed that the testimonies of unassailable gentlemen support experimental results. In a letter sent in March 1666 to Henry Stubbe (who had dedicated his 1666 *Miraculous Conformist* to Boyle), Boyle applied his arguments: "I am very back-warded to believe any strange thing in particular, though but purely natural, unless the testimonies that recommend it be proportional to the extraordinariness of the things proposed: yet I remember not, that I have hitherto met with any, at least any cogent, proof that miracles were to cease with the Age of the Apostles and not only the excellent Grotius."[36]

Boyle found an agreeable approach to the independent construction of knowl-

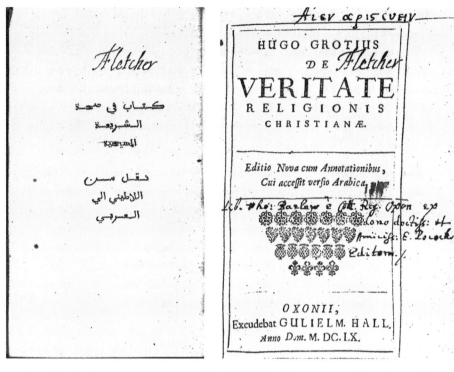

On *right*, the front cover, in Latin, and on *left*, the back cover, in Arabic, of Pococke's translation of Hugo Grotius, *De veritate religionis Christianae*. Instead of fronting pages, Latin text began at the front cover and the Arabic translation at the back cover. *De veritate religionis Christianae: Kitāb fī ṣaḥat al-Sharīa' al-Mashīhyyah*, translated by Edward Pococke (1660). Courtesy of the Clark Library, Los Angeles.

edge in Grotius's *De veritate*. He appropriated Grotius's argument about the necessity of trustworthy testimonies for belief in miracles and applied it to firsthand experience and self-reliant learning. In the same vein, he gave credence to his own air pump experiments: since all experiments cannot be replicated, results based on the testimonies and consensus of creditable gentlemen should be trusted.[37]

The Aspiring Naturalist

Already during the 1660s, before the publication of *Philosophus autodidactus*, Pococke had familiarized Boyle with the story of Ḥayy. In a letter from January 1660, Samuel Hartlib mentioned that Boyle was considering sponsoring the publication of Pococke's "Ingenuous Arabic Fiction." As a fervent follower of Bacon, Hartlib was excited about the story of Ḥayy and committed himself to further convincing Boyle to sponsor its publication.[38]

Boyle's interest in the "ingenious Arabick fiction" went beyond the actual maintenance of personal network. The very notion of a resourceful prodigy, Ḥayy, who experiments with nature in complete seclusion and celibacy on a deserted island resonated in Boyle's social world, even in personal matters of love and domesticity.[39]

The trope of a child prodigy who travels and explores world and nature had already manifested in his early writings. He had composed an autobiographical work about his childhood, written in the third person, in the late 1640s. *An Account of Philaretus During His Minority* reflects Boyle's concern for the centrality of self-directed knowledge and character formation, subjects also addressed in his moral essays. He organized the piece around the pursuit of happiness, reflections on which recur at intervals throughout the text. Similarly, his discussions of morality, with their Christianized Stoicism and ideals concerning self-control and the triumph of internalized virtue against external forces, recur.[40] Boyle, in the guise of the prodigy Philaretus, relates that he had been separated from his parents and sent away from home, committed by his father to the care of "a Country-nurse" who "early inur[ed] him by slow degrees to a Course [of practical living]."[41] Pococke was familiar with Boyle's autobiography,[42] and the image of Philaretus resonates with that of Ḥayy Ibn-Yaqẓān. Both practiced trial and error in the service of reason and experienced solitude and detachment from parents. Both pursued happiness through self-control and, as mentioned, strength in the face of external forces.

Boyle also ventured into scientific utopia writing. During the 1660s, he had started writing "a short Romantick story," though he was able to complete only two pages of his utopia, "The Aspiring Naturalist." His hero, Authades ("self-willed"), who lives on a utopian island in the "southern ocean," narrates to a European visitor, Philaretus, his discoveries in natural philosophy. The self-learning of Authades counterpoised European authoritative knowledge with the inhabitants' self-exploration of nature. In Europe they straiten the "soul by narrow opinions and prejudices of learned men . . . who are not acquainted with anything in the Art of nature." Since those learned man "are not of free and emancipated genius, and are not accustom to see and to make reflections of the wonders of nature, they cannot but be inclined to estimate the possibility of familiar natural phenomena." "The philosophers of our Island," however, "are able to perform" experiments, invoking hidden phenomena and making discoveries that might in Europe be seen as extravagant.[43]

Mechanical Spontaneous Generation

In 1660 an account of Robert Boyle's work with the air pump was published under the title *New Experiments Physico-Mechanicall, Touching the Spring of the Air, and its Effects*. It was while answering the objections of his foes that Boyle made first mention of the law that the volume of a gas varies inversely to the pressure of the gas, a law that is usually called after his name. The work was published during the 1660s in various editions, and in 1669 it left England and was first published in Latin in Rotterdam.

Philosophus autodidactus furnished Boyle with cultural symbols and philosophical arguments, allowing him to further promote the experimentalist program. The heated exchanges over Boyle's air-pump experiments, along the second half of the 1660s, made the translation of *Ḥayy Ibn-Yaqẓān* pertinent not merely to the autodidactic methodology but also to the connection of Ibn-Tufayl to two major questions in natural philosophy: motion in a void and spontaneous generation. For medieval Andalusian philosophers, Avempace—and his follower Ibn-Tufayl—natural motion can be measured by subtracting the resistance of a medium through which an object moves. As a result, various densities of media would result in various velocities.[44] Ernest Moody connected Avempace's arguments for the possibility of motion in a void with medieval scholastic scholars and with Galileo's experiment and arguments about falling bodies.[45] Moody noted that Boyle, familiar with this tradition, took the argument one step further, becoming the first to experiment with the possibility of motion in a vacuum by creating an artificial vacuum in an air pump, eventually showing that the volume of a gas rises as the pressure of the gas falls.

In the note to the reader in *Philosophus autodidactus,* Pococke links Ibn-Tufayl to Avempace and refers to his experimental method, "which after a number of centuries can perhaps still seem new."[46] The general ignorance about Ibn-Tufayl, as well as the intriguing connection to Avempace, interested early modern scholars, who at the time knew far more about Avempace than they did about Ibn-Tufayl.

Spontaneous generation further connected *Ḥayy* to Boyle. In the late 1650s, when Pococke and Boyle started their exchanges, Boyle wrote his *Essay on Spontaneous Generation*. Extant passages show that Boyle discussed the possibility of spontaneous generation and offered a mechanical explanation in which, in

Frontispiece, Robert Boyle, *Nova experimenta physic-mechanica de vi aëris elastica & ejusdem effectibus* (Rotterdam, 1669), showing scholars experimenting and self-learning. Courtesy of the Houghton Library, Harvard University.

certain circumstances, the sun provokes generation out of existing matter, just as "we see that when the sunbeams are suddenly started upon our head they oftentimes excite and determine the spirits in the brain and some other parts after the manner requisite to produce that motion and sound which we call sneezing."[47] From the description of sneezing, Boyle works to establish a mechanical explanation for spontaneous generation. The creation of the first individuals of each species, from which life evolved, was, as in Pico's description, dependent on godly mechanical laws—on particular relations between the location of the sun on the horizon and the rare mixture of matter. For Boyle, the conditions that allow or prevent spontaneous generation have all been calculated and included in the grand design of the "Author of the Book of Nature."[48] A rare combination of certain protoplasts in which the sunbeams might provoke change was, therefore, necessary for spontaneous generation to occur. Even more, a balanced environment, where the various organisms were in a particular proportion and the climate was neither too hot nor cold, had to be present.[49]

The question of spontaneous generation, not considering its apparent contradiction with monotheist creationism or with Aristotelian eternality, scarcely existed in the years between Pyro, Epicurus, and Lucretius and Boyle's era. In the seventeenth century a few sources reviewed the question in different contexts.[50] In most cases it came up in relation to the biblical story of the creation of Adam. One conspicuous medieval source for spontaneous generation did, however, exist: the story of Ḥayy's creation on a desert island, beyond the horizons of climate and civilization, where the natural conditions of humidity and balanced weather combined with an endowment of supernal light. All these conditions came together to create the perfect conditions for the spontaneous generation of human beings.

Spontaneous generation remained a difficult concept in Pococke's time, and in his note to the reader he presents the intellectual sources of the concept that Ibn-Tufayl may have used. The pious Pococke, however, tried to lighten the argument in favor of spontaneous generation in observing that Ibn-Tufayl "seems to have acted less carefully than those who are accustomed to relate fictive stories, while he relates those things that happen to an infant either sprung from the earth on that uncultivated island or taken somewhere else not long after its birth, he attributes everything to the report and to the confidence of many men who are not the sort to be lacking in knowledge, and who know the island of Wāqwāq, abundant in marvels, just as well as they know their own homes."[51]

Independently Learned Natural Religion

Philosophus autodidactus reaffirmed Boyle's arguments regarding experimentation and the possibility of spontaneous generation, but it also presented a unique way of experimenting with and exploring nature, activities that could lay a foundation not only for natural philosophy but for Unitarian theology as well. In 1690, twenty years after the publication of *Philosophus autodidactus*, Boyle published

The Christian Virtuoso: Shewing that by Being Addicted to Experimental Philosophy a Man is Rather Assisted than Indisposed, to be a Good Christian, in which he applied experimental philosophy to Christianity.⁵² However, as fragmentary documents indicate, Boyle started composing it early in 1670.⁵³ Through the 1680s he revised the work, incorporating notes that John Locke suggested after reading the draft⁵⁴ and finally publishing it before his death (in 1691) as a program for the self-taught Christian.

The Christian Virtuoso shows that experimental philosophy, regardless of its manipulation of nature and seeming disregard of first principles, can and should be applied to religion. He stressed the need for experience and practice in building up knowledge not only of nature, but also of theology and God: "The virtuosi I speak of... consult experience both frequently and heedfully; and, not content with the *Phaenomena* that nature spontaneously affords them." In soliciting nature through trial and error, Boyle's virtuosi "have a peculiar right to the distinguishing Title that is often given them, of *Experimental Philosophers*."⁵⁵

Boyle saw both a philosophical and cultural need to write this work. The *Christian Virtuoso* presented the essence of experimentalism as actual acts of communion with nature, which then led to communion with God. The new mechanical philosophy, with its strong emphasis on matter and man's perception, raised secular arguments affiliated with atheist atomism.⁵⁶ Boyle's attack on atheist atomism, therefore, aimed to purify experimental philosophy of any suspicions and show its advantages in a religious context. For Boyle, experimentalism set the foundation for the exploration of natural religion and the means to read the book of God against the book of Nature. To achieve the knowledge of God as the author of the book of nature, man has to rely not only on reason and contemplation but also on practical experience of nature.⁵⁷

In addition to its contemporary religious and political appeal, *Philosophus autodidactus* drew the interest of religious thinkers in a way that Ibn-Tufayl himself had always hoped: the text promised a method of attaining a mystical relationship with God. A clue to its mystical and illuminist aspects shows up in the first English translation (1671) of *Ḥayy Ibn-Yaqẓān*, made from Pococke's Latin. The translator, George Keith, a prolific and controversial Quaker who published anonymously, wished to support the argument that religion is a matter of intense personal experience and that one's inner light can eventually secure knowledge of God.⁵⁸ He titled his translation *An Account of the Oriental Philosophy*, joining the notion of oriental philosophy with that of illumination. Keith writes in the preface,

After it came into my hands, and that I perused it, I found a great freedom in mind to put it into English for a more general service, as believing it might be profitable unto many; but my particular motives which engaged me hereunto are to show that he [Ibn-Tufayl] illustrated excellently how far the knowledge of a man, whose eyes are spiritually opened, different from that knowledge that men acquire simply by hear-say, or reading: and what he speaks of as a degree of knowledge attainable, that is not by premises and conclusions deduced but by firsthand practical experience that brings him into conjunction with the supreme intellect and with a certain truth.[59]

Keith's interest in Pococke's work seems to have derived from *Hayy Ibn-Yaqzān*'s epistemological proposal that knowledge should not stem from oral and written traditions but should instead rely on "native inward testimonies of the mind." Moreover, the conjunction of the experimentalist and mystical illuminist aspects of practice fit the Quaker's preference for experimentalist science. The number of Quakers in the Royal Society exceeded that of any other religious group, mostly for the primacy they attributed to direct experience over scriptural authority.[60]

The religious reception of *Hayy Ibn-Yaqzān*, as it appeared either in the *Christian Virtuoso* or Keith's translation into English, exemplified the broader problem of knowledge formation and the roles that natural philosophy and theology play in it. But the Quakers were not the only people drawn to Pococke's publication for sound philosophical and religious reasons.

Hayy Ibn-Yaqzān and the Blank Slate

Pococke was also a member of a network of leading Oxford experimentalists closely connected with John Locke, the founder of empiricist philosophy. While learning mathematics and astronomy under John Wallis and Seth Ward during the 1650s and early 1660s, Locke also attended Pococke's twice-weekly classes in Arabic and Hebrew. In 1663 Locke applied for a tutorial position and received endorsement through a statement of recommendation signed by three professors, including Pococke.[61] After *Philosophus autodidactus* was published, Locke intensified his interest in the original cultures of Near Eastern philosophical text objects.[62]

Only a few bilingual publications could be used as textbooks in the learning sessions that occurred between Pococke and Locke. Pococke published one of them—a commentary on the Mishna by Maimonides—in 1650, under the title *Porta Mosis*. It was laid out in two columns, one containing the original Judeo-

Arabic and the other a Latin translation. This may have been where Locke encountered the statement by first-century scholar Elisha Ben-Avuyya, which appears in chapter 4 of the Mishna, "To teach a little boy is just like writing on a blank slate, and to teach an old man is just like writing on an erased slate."[63] Later that year, Locke's fascination with Maimonides continued, and he employed Pococke's son in a project to translate Maimonides' *Guide for the Perplexed*.[64]

But the importance that Pococke and Locke played in one another's philosophical lives lasted well beyond the years they spent studying together, and when Pococke lay dying in his bed, his masterpiece *Philosophus autodidactus*, the story of the autodidact and ultimate *tabula rasa*, made its last and most impressive imprint. Locke had already started working on what would become a canon of empiricism, *The Essay Concerning Human Understanding*, in 1671 but published it only in 1689, calling all men to "think and know for themselves." In the first book of the *Essay*, Locke uses quintessential exemplars to refute the possibility of innate ideas. "Children, idiots, savages, and illiterate people," he observes, are the "least corrupted by custom, or borrowed opinions" and "have received the least impression from foreign opinions." One who would expect to find abstract maxims and reputed principles of science in "a child untaught, or a wild inhabitant of the woods," Locke suggests, "will find himself mistaken."[65]

How, then, do men commonly come by their principles? "In that which every day's experience confirms." To exemplify, Locke asks his readers to consider the ways and steps by which "doctrines have been derived from the superstition of a nurse, or the authority of an old woman . . . For such, who are careful to principle children well, instil into the unwary, and as yet unprejudiced, understanding, (for white paper receives any characters,) those doctrines they would have them retain and profess." Ideas, especially those belonging to principles, are not born with children: "If we will attentively consider new-born children, we shall have little reason to think that they bring many ideas into the world with them."[66]

Exploration of nature that leads independently to a communion with God, Locke suggests, refutes the argument that the idea of God is innate. "The notion of a God," Locke proposed, might carry credence "were it to be found universally in all the tribes of mankind, and generally acknowledged, by men grown to maturity in all countries. For the generality of the acknowledging of a God, as I imagine, is extended no further than that; which, if it be sufficient to prove the idea of God innate, will as well prove the idea of fire innate; since I think it may be truly said, that there is not a person in the world who has a notion of a God, who

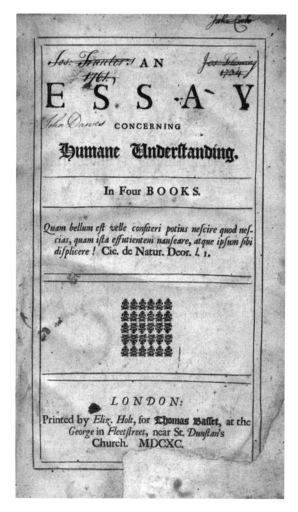

Title page, John Locke's *Essay Concerning Human Understanding*, 1st edition (1690). Courtesy of the Houghton Library, Harvard University.

has not also the idea of fire." Locke takes his argument a step further, exemplifying it through a case study:

> I doubt not but if a colony of young children should be placed in an island where no fire was, they would certainly neither have any notion of such a thing, nor name for it, how generally soever it were received and known in all the world besides; and perhaps too their apprehensions would be as far removed from any name, or notion, of a God, till someone amongst them had employed

his thoughts to inquire in to the constitution and causes of things, which would easily lead him to the notion of a God; which having once taught to others, reason, and the natural propensity of their own thoughts, would afterwards propagate, and continue amongst them.[67]

In 1690 Locke sent to his teacher, Pococke, a copy of his essay with a dedication in which he expressed his gratitude and appreciation for Pococke's teaching and inspiration. The dying teacher, however, could not respond in writing; as his son later wrote to Locke, his father's "age and infirmities" kept him from "delivering his words that he heartily valued" Locke's present.[68]

The close connection between Locke and his teacher climaxed less than a year later (1691) when Locke was chosen to write the eulogy for Pococke's funeral memorial. Describing the effects of the Civil War on Pococke's disposition, Locke said, "Indeed he was not forward to talk, nor ever would be the leading man in the discourse . . . He would often content himself to sit still and hear others debate in matters which he himself was more a master."[69]

Locke, apparently, provides testimony that Pococke, the only royalist left standing, who had witnessed his close friends' grim ends, especially that of John Greaves, ended up subdued, functioning under a certain level of persecution, especially during the 1650s and 1660s. His precarious social and political standing compelled him to protect himself from further unpleasant turns and to manipulate his public image by publishing works like *Philosophus autodidactus*. Locke's eulogy further supports this view of a subdued Pococke by claiming that "at the Restoration of King Charles [II] [Pococke's] merits were so overlooked or forgotten, that he was barely restored to what was his before, without receiving any new preferment then or at any time after."[70]

Philosophus autodidactus also helped form Locke's notions of education. In 1693, three years after writing the eulogy for his admired teacher, Locke published *Some Thoughts about Education*, his correspondence with Edward Clark regarding the education of Clark's son. In his dedication to Clark, Locke testifies, "I myself have been consulted of late by so many, who profess themselves at a loss how to breed their children, and the early corruption of youth is now become so general a complaint . . . The method here proposed . . . examined and distinguished what fancy, custom, or reason advises in the case, set his helping hand to promote everywhere that way of training up youth, with regard to their several conditions, which is the easiest, shortest, and likeliest to produce virtuous, useful, and able men."[71]

In this piece, which became a pillar of liberal education during the Enlightenment, Locke explains how to educate a *tabula rasa* using three distinct methods: the development of a healthy body; the formation of a virtuous character; and the choice of an appropriate academic curriculum. These methods had a broader appeal since his educational principles allowed women and the lower classes to aspire to the same kind of character as the aristocrats for whom Locke originally intended the work. *Ḥayy Ibn-Yaqẓān* raised the possibility of learning without any guidance and teachers, and then Locke came along and propagated it more widely. If a spontaneously generated wild boy was able to rise to the most divine philosophical questions through a course of independent study, then with the aid of education, anyone, even a boy from a poor family, could become a useful citizen. The autodidact became a role model for new educational programs with universal claims.

As the case of Pococke's *Philosophus autodidactus* shows, a process of exchange and borrowing from sources both near and far combined to construct the edifices of experimentalism and empiricism. The economic network of the Levant Company, for example, allowed for the circulation of cultural and scientific works that, as oracles of the past, played roles in reinforcing the experimentalist argument still being negotiated, before the canon of empiricism solidified in the form of Locke's *Essay Concerning Human Understanding*. Despite his ulterior motives, Pococke's *Philosophus autodidactus*, with its tropes of wilderness, solitude, and autodidacticism, played its own role in that negotiation.

The investigation of Pococke's networks has shown that his collecting and translation of works from the Near East, as well as his struggle to obtain and keep his chair at Oxford, not only reflected the agendas and chronology of his own career struggles but also mirrored his relationships with leading experimentalists. Since the controversies over experimental philosophy arose after he returned to Oxford, it seems that he first picked up Ibn-Tufayl's *Ḥayy Ibn-Yaqẓān* because it appealed to his personal taste, and only later, when experimentalism represented the prevailing trend, did he grasp the opportunity and decide to publish it.

This particular medieval work appealed to English experimentalists because it carried within it traces of earlier intellectual and cultural ideas about empiricism, which ranged from the question of self-reliance and self-directed learning to spontaneous generation. The affair of the publication of *Philosophus autodidactus* also tapped into an already extant history of a utopian literary genre with sources in medieval times. Remote and imaginary islands had been used in medieval intellectual culture to describe extraordinary places representing the birthplace of the laws of nature and of civilization. These islands stood as laboratories where philosophers could imagine, openly and without fear of persecution, the spontaneous generation of monsters, exceptional creatures, and even human beings capable of knowing nature and God without the burden of traditional authority. Beginning in the sixteenth century, philosophers had used this genre to popularize critiques of prevailing intellectual and cultural discourses.

CONCLUSION

Sampling the History of Autodidacticism

IN CONTEMPORARY SCHOLARSHIP and public discourse, the meaning of *autodidacticism* extends beyond self-directed learning. Enthusiasm for unguided education and a high degree of self-motivation, though, also points to the modern faith in the individual's ability to disregard institutions and control his own destiny. In many ways autodidacticism can be extracted from the history of experimental natural philosophy, republican political philosophy, and liberal ideologies, each of which holds up self-teaching as a primary agent for the transformation of society, from the bottom up. In modern science, experimentalism glorifies man as an industrious virtuoso; through technological advances, he projects his mechanical manipulation of nature onto society, controlling nature to extract the necessary means of his subsistence. Within political philosophy, republicanism exalts human beings as rational creatures endowed with goodwill, who can access the values of the public will and the common good without recourse to superimposed authority, thereby creating a government of the many. In the realm of education, liberal ideologies praise self-cultivation as the central agency of social mobilization and political transformation.

Autodidacticism is most closely identified with an Enlightenment ethos, which Immanuel Kant compressed to two words of Horace—*Sapere aude,* "Dare to know." Early modern utopias, however, had already used the tenets of autodidacticism in describing the bottom-up structure of perfect societies, which depended upon firsthand experience guided by individual rational thought. Although most of these utopias required education systems, conformity was generated not by primordial principles or institutionalized authority but by common sense grounded in personal experience.

Historiography presents the early modern utopian tradition as a reason-based emulation of ancient Greek utopian tales, such as Homer's "Isles of the Blest"; Plato's description of the happy life in Atlantis, gone forever as a result of some unknown natural disaster; and Virgil's *Saturnia regna* (The kingdom of Saturn), in which "a new breed of men, sent down from heaven," return justice to the world.¹ Utopian thought also looked to the Judeo-Christian Bible, especially its story of an earthly paradise in which Adam and Eve were commanded to assign names to natural objects, thereby taking control of nature.

Isaiah Berlin further strengthened the connection between early modern utopias and autodidacticism, arguing that at the heart of both Greek and early modern utopian visions lay the Platonic assumption that virtue is knowledge: "If you know the good for man, you cannot, if you are a rational being, live in any way other than that whereby fulfillment is that towards which all desires, hopes, prayers, aspirations are directed." The desertion of this Platonic virtue, Berlin judged, led to a distinct decline in the appearance of utopias during the medieval period, "perhaps because according to the Christian faith man cannot achieve perfection by his own unaided efforts; divine grace alone can save him." He further argued, however, that the same Platonic assumption, sometimes in its baptized, Christian form, eventually animated the great utopias of the Renaissance.² For Berlin and other scholars, medieval culture represents the dark ages of the utopian tradition, separating the Greek tradition from the early modern. Only in the seventeenth century could utopias like those created in Thomas More's *Utopia* (1516), Tommaso Campanella's *City of the Sun* (1602), Francis Bacon's *New Atlantis* (1623), and more, come to the fore, calling upon their readers to place their trust in no book, no teacher, and no traditional authority but to rely solely on firsthand experience, on self-directed exploration of nature.

An important geocultural proposition implied that utopias represented Western values—projecting the Greek tradition onto the New World, onto the future. Early modern utopias have been conceived by Berlin and other eminent intellectuals as marking a radical shift away from medieval perceptions of man as subjected to animated nature and its spiritual influences in favor of a modern, European perception of the industrious man who conquers a passive nature and thus takes control of his destiny.

The shift from the medieval intellectual culture guided by *vita contemplativa*, focused on the relations of man with God, to the early modern call for *vita activa*, preoccupied with the relations of man with the cosmos, represents a break in space. The rise of experimentalism, republicanism, and liberalism happened in Europe, representing topographical features in Europe's historical development as it sought to break from adjacent cultures still caught up in the human impotence of *vita contemplativa*. While current historiography conceives of Europe as the locus of modern science and of the aspiration to explore and control nature, it views the Near East as spiritually subjected to an animated nature that followed intellectual traditions faithfully transmitted through many generations. To be sure, this lasting historiography has stressed that autodidacticism and scientific utopias stand as distinct early modern European constructions. The West has been presented as the only rebellious civilization to have shucked off the shackles of tradition.

Autodidacticism, however, is not a modern European novelty. Between the utopias of the Greeks and those of the early moderns, an autodidactic utopian vision, *Ḥayy Ibn-Yaqẓān*, stimulated various controversies over autodidacticism. In exploring the twilight zone of that temporal and spatial break, this historical sampling of autodidacticism implies an overarching, often misinterpreted cultural space hovering over the intellectual cultures of Marrakesh, Barcelona, Florence, and London.

Sampling the various cultural contexts within which single utopian treatises circulated links major European humanists, utopians, and experimentalists to *Ḥayy Ibn-Yaqẓān*. A subterranean historical current tied medieval Near Eastern mystical philosophers to utopian writers and experimentalists who were engaged with the same text. All shared a concern with the question of thought and experience and all promoted new practices that would lead to the attainment of truth. Each had a particular interpretation. While the mystics promoted intuitive perception and communion with nature and God, the utopians and experimentalists promoted new mechanical epistemologies based on sensations and impressions that came from firsthand experience.

An examination of the textual objects and diverse circumstances within which they circulated allows for a reevaluation of our understanding of the goal of early modern autodidacticism. Rather than the modern perception of self-discovery by way of pure reason, with the eventual goal of secularism, early modern humanists perceived self-discovery as a teleological process based on intuition, moving the individual toward the goal of communion with nature and God and thereby to the ultimate felicity. Philosophers imagined firsthand experience as a combination of practice and theory that lauds the making of philosophy without the technique of philosophy. The premises that lay the foundations for the early modern self—experimentalist, republican, and liberal—-resonate with medieval sources, stressing the need to interact with nature and God and not just contemplate them.

In the four freestanding historical moments in which *Ḥayy Ibn-Yaqẓān* was embedded, autodidacticism correlated with particular cultural concerns, beginning with an Andalusian attempt to cope with the challenge of the science of practice of Sufism; Spanish-Jewish pedagogical considerations about the peril of teaching philosophy to boys; a Florentine struggle over sodomy and astrology; and the rising experimentalist culture of the seventeenth century. The description of the various readings of a single text in a range of contexts allows those disparate cultural nexuses to communicate with one another and to give way to an alternative history, interdisciplinarily fusing previously excluded late medieval European and Islamic-Jewish intellectual cultures.

Faced with cultural challenges, Ibn-Tufayl turned to the cultural and philosophical sources at his disposal—wonder stories of the Far East, mystical tradition from the Arab East, and local references to prehistoric man—to create a philosophical utopian novel. He used the story to convey a radical rejection of tradition, proposing firsthand experience and the ideal of autodidacticism in its stead. The book's role as ethnologic satire allowed for the introduction of bold philosophical arguments concerning spontaneous generation, the ability to acquire knowledge without authority, and the life of solitude. Situated on the mythological island of Wāqwāq, which lay beneath the horizons of climates and civilizations, the regular laws of nature and society did not apply. Despite its radical premise, though, Ibn-Tufayl reconciled the ongoing controversy over al-Ghazzālī's *Revival*, intermingling the Sufi science of practice with the Andalusian logic-based philosophical tradition. In appropriating the challenging Sufi science of practice into a system of gradual unguided education in philosophical principles, he moved the emphasis from the contemplative to the active exploration of nature.

Ḥayy Ibn-Yaqẓān was translated into Hebrew sometime in the thirteenth century, part of the process that took Arabic philosophical texts from Spain, through Catalonia, across the Pyrenees Mountains to Provence, producing a wave of translations, particularly of Averroeist writings, into Hebrew and Latin. These translations sparked local controversies between philosophers and theologians among Christians at the University of Paris and Jews in the communities of Provence-Catalonia. More than a century later, in 1348, Moses Narbonni arrived in Barcelona. He set out to write a commentary on the anonymous Hebrew translation of *Ḥayy Ibn-Yaqẓān*, producing his *Yehiel Ben-'Uriel*.

At the time of Narbonni's birth, in the late thirteenth century, teachers introduced local boys to philosophy; the boys also had access to philosophical speeches delivered on public occasions. That permissive study of philosophy, however, led to strong objections by leading conservative scholars from Provence and Catalonia. They struggled to limit philosophy's access to older, educated men and issued an excommunication ruling that forbade the teaching of philosophy to young boys. Leading scholars in Perpignan reacted by arguing for and justifying the study of philosophy.

No better text existed to defy the excommunication ruling and promote autodidacticism than *Ḥayy Ibn-Yaqẓān* the story of a boy who derived philosophical principles through trial and error and exploration of nature. In *Yehiel Ben-'Uriel*, Narbonni not only defended the pedagogical argument that philosophy could be taught to younger boys but also supplied his community of Perpignan with additional evidence in favor of autodidacticism, sudden perception, and the experimental exploration of nature. Interest in autodidacticism grew in the wake of Narbonni's commentary. Readers came to know *Ḥayy Ibn-Yaqẓān* through his work in Hebrew and Latin. During the fifteenth century, manuscripts of Narbonni's commentary circulated through the Jewish communities of Barcelona, Provence, and Italy. During the 1480s, Jochanan Alemanno, whose own manuscript of *Yehiel Ben-'Uriel* contains extensive marginalia, went on to become the key figure in the translation and transmission of *Yehiel Ben-'Uriel* into Latin.

In the early 1490s, Giovanni Pico della Mirandola encountered one of these circulating manuscripts and had it translated into Latin for the first time. Pico, a wild prodigy himself who took every opportunity to defy authority and social convention, met Jochanan Alemanno in Florence in 1488. Alemanno introduced him to *Ḥayy Ibn-Yaqẓān* while Pico was completing the *Heptaplus*, where he first mentions Ibn-Tufayl and first trains his eye on the biblical Adam as the original autodidact who directly explored and conquered nature.

In Florence, Pico experienced the interplay between erotic love and the ideals of platonic love. After the death of his patron, Lorenzo de' Medici, in 1492, different astrological accounts predicted the destruction of Florence, which, like the biblical Sodom, would go down because of its permissive attitude toward sodomy and homosexuality. As a response to attitude of Lorenzo's time, when adolescent boys played a central role in the homosexual Florentine street culture, once Savonarola took control of the city, both those boys and the men who engaged with them were subjected to persecution. Disturbed by the astrological predictions, Pico cloistered himself in his suburban villa and wrote his screed against astrology, among his sources being a Latin manuscript-translation of *Ḥayy Ibn-Yaqẓān* In Ḥayy's story he found corroboration for his ideas about natural magic as the practice of natural philosophy and as a form of Platonic love. He used the Andalusian novel to show that firsthand experience molds the lives of young boys.

Although never printed, Pico's translation of *Ḥayy Ibn-Yaqẓān* soon made an impact on local Renaissance culture. Antonio Fregoso (d. 1515), a contemporary

Italian poet, read it and went on to transform the philosophical novel into a poem, *De lo istinto natural,* which praises natural instinct as the principle guide for the autodidactic life.

Giambattista della Porta, in his *Magia naturalis* (1558), had further developed Pico's perception of natural magic as the self-learning practical method to explore nature, laying out a list of experiments and observations on the natural world. He set out his self-learning methodology in describing the ways "the knowledge of secrecies depends upon the survey and viewing of the whole World," stating that the knowledge of secret things depends upon "the contemplation and view of the face of the whole world, namely, of the motion, state and fashion thereof, as also of the springing up, the growing and decaying of things. For a diligent searcher of *Nature's* works, as he sees how nature does generate and corrupt all things, so does he also learn *Physic, Husbandry,* the art of building, the disposing of household affairs, and almost all arts and *Sciences.*"[3]

Della Porta took his methodology a step further, employing it as a studying program for his close circle. In 1580 he founded a scientific society called the *Academia Secretorum Naturae* (Accademia dei Segreti), more commonly known as the Otiosi (Men of leisure), which aimed to study the secrets of nature. Any person applying for membership had to demonstrate that he had made a new discovery in the *natural sciences*. Pope Paul V, however, disbanded this utopian academy in 1578.

In early modern England, self-teaching resonated in the emerging utopian tradition. In 1510 Thomas More translated the *Vita de Pico,* Gianfrancesco Pico della Mirandola's biography of his famous uncle. We know that More read Pico's *Opera,* where Ḥayy's story is mentioned as a treatise showing "how anybody can become a philosopher." But before he began to write *Utopia,* More was already engaged with other sources, satirically criticizing the authoritative modes of learning. In 1509 Erasmus of Rotterdam dedicated to More a satirical oratory, *The Praise of the Folly,* after the manner of the Greek satirist *Lucian,* whose work Erasmus and More had recently translated into Latin. In a series of ironic orations Erasmus further marked the transition in icons of wisdom from the old man to the child. *Praise of Folly* ironically identifies wisdom (*sofia*) with doctrines, institutions, and old men and folly (*moria*) with fresh perception, happiness, and children. "For who would not look upon that child as a prodigy that should have as much wisdom as a man?" Erasmus satirically asked. "Or who would endure a converse or friendship with that old man who to so large an experience of things had joined an equal strength of mind and sharpness of judgment?"

In the following years, More became preoccupied with mystical practice and the writing of poetry, culminating in the publication of his *Utopia*. Set on a remote island, the inhabitants have replaced *vita contemplativa* with *vita activa* and improved their natural abilities through self-directed exploration guided by reason and thus have established a government of the many based on collective goodwill.

Rather than basing his utopian vision on the transmission of ideas or the superimposition of metaphysical premises, More imagined its philosophical underpinnings as growing from the bottom up. Children and education turn out to be central to that vision, with the idea that in the Utopia, the adults "instill in children's minds, while they are still tender and pliable, principles useful to the commonwealth."

More further presents the Utopians lively exploration of nature. "They reckon the knowledge of [physics] one to the pleasantest and most profitable parts of philosophy, by which, as they search into the secrets of nature, so they not only find this study highly agreeable, but think that such inquiries are very acceptable to the Author of nature." More then asks his reader to imagine that God, "like the inventor of curious engines among mankind, has exposed this great machine of the universe to the view of the only creatures capable of contemplating it" and suggests that the curious observer, who admires the work of God, "is much more acceptable to God than one of the herd, who, like a beast incapable of reason, looks on this glorious scene with the eyes of a dull and unconcerned spectator."

Finally, in the only sample of the Utopian language that he provides, More makes a striking statement lauding the self-directed exploration: "The commander Utopus," he writes, "made me, who was once not an island, into an island. I alone of all nations, without philosophy, have portrayed for mortals the philosophical city." More further extended the plea to autodidacticism by calling on sixteenth-century humanists to make philosophy out of firsthand experience, without the prerequisite metaphysical foundations.

More was the first European writer to employ the parable of the island for the benefit of utopian societies that pursue happiness through independent learning and self-discovery. Subsequent generations followed his lead: Tommaso Campanella's *City of the Sun* and Francis Bacon's *New Atlantis* transformed this political program for active self-discovery into revolutionary programs of natural philosophy. They presented the pursuit of happiness as a gradual discovery of the self through the exploration of nature. To be sure, in exploring nature and internalizing its mechanical manipulations, isolated utopian people on far-flung islands actually recreated themselves as industrious virtuosos who invented

Sampling the History of Autodidacticism   133

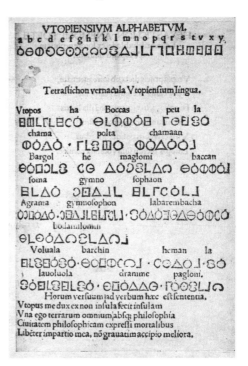

On *left,* aerial view of the island Utopia, surrounded by waters with two sailing vessels; standing in *left foreground,* Morus and Hythlodaeus; *right foreground,* Petrus Aegidius. On *right,* the Utopian language, with Latin translation in the paragraph below, stresses the autodidacticism of the Utopian people who make philosophy without philosophy. Thomas More, "De optimo reip. statu deque nova insula Utopia...." Engraved and printed by Johann Froben (Basil, March 1518). Courtesy of the Houghton Library, Harvard University.

technological links between the theoretical dimension of science and the practical needs of society.

Despite the similarities, the locations of these utopian island settings varied from one text to another. Ḥayy's island, Wāqwāq, in the Indian Ocean, situated autodidacticism close to its imaginary origins, the oriental philosophy of the Indian Ocean region, which at the time represented the ends of the earth. For Thomas More, the island lay somewhere in the Atlantic Ocean, in the New World. Campanella's City of the Sun is also in the Indian Ocean, on the mythical island of Taprobane, near Serendib (Sri Lanka), where one can find "Adam's footprint after his fall from grace." Campanella tells his readers that "this race of men . . . came there from India, flying from the sword of the Magi." The people of the city follow Brahma and Pythagoras, and the necessary values to make society work by itself are imprinted on children's minds, thus achieving a political order built from the bottom up. But political order is also achieved in reflecting on the natural order.

When children reach six years of age, the Solaris of Taprobane teach them natural philosophy and the mechanical science. They pick children who would further study nature by releasing them into wild nature and consider them the more noble and renowned children, "who ha[ve] dedicated [themselves] to the study of the most arts and [know] how to practice them wisely." The knowledge of politics and of nature are essential in the ruler, "since he who is suited for only one science and has gathered his knowledge from books, is unlearned and unskilled." But this is not the case with the Solaris' children, who possess "intellects prompt and expert in every branch of knowledge and suitable for the consideration of natural objects" and who study the "sciences with a facility by which more scholars are turned out by us in one year" than by other nations "in ten, or even fifteen."

Campanella's connection to two figures transformed the utopian autodidactic vision into an actual program for a self-learning experimental society. Before he was imprisoned in Naples, he was associated with his friend Giambattista della Porta's *Academia Secretorum Naturae*. In 1603, Campanella and della Porta inspired Federico Cesi and his friends to establish the first academy of the sciences, the Accademia dei Lincei. Cesi selected the name to symbolize the desire to see into the secrets of nature with a perception as acute as that of the lynx. The self-taught spectator of natural philosophy, equipped "with lynx like eyes, would examine those things which manifest themselves, so that having observed them, he may zealously use them."[4] Through free experiment, unshackled by blind obedience to the authority of Aristotle and Ptolemy, they sought to establish a new science based on firsthand experience that would call the philosophical tradition into question.

The society soon left marks on scientific discoveries. As the seventh Lincei, Galileo sought Cesi's help in publishing his discoveries, since he ran into censorship problems with his *History and Demonstrations Concerning Sunspots and their Phenomena*. Cesi suggested publication of the sunspot observations in the form of letters, headed with an epigraph of Horace: "Courage opens wide the gates of heaven."[5] With Horace at the mast, the sunspot letters were published in 1613 under the imprint of the Linceian Academy and soon brought the question of the earth's motion and its compatibility with the scriptures to the fore.

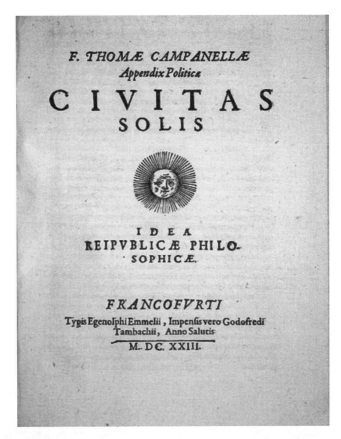

Frontispiece, Tommaso Campanella, *Civitas solis: idea reipublicae philosophicae* (Frankfurt, 1623). Courtesy of the Houghton Library, Harvard University.

Francis Bacon carried on the Linceis' vocation and called on scholars to admit nothing but on the faith of eyes. He took up the idea of the island utopia and, looking to the New World, set it off the coast of Peru, in the Pacific Ocean. Bacon's *New Atlantis* places autodidacticism at the point of interface between rational experimental practice leading to union with nature and mystical practice leading to communion with God.[6]

More than any other scholar of his time, Bacon called for the reliance on firsthand experience, turning to autodidactic exploration to reveal the secrets of nature, improve life, and achieve communion with God. Similar to Pico in his view of nature, he described nature as an inanimate organ that needs to be explored through practice. The "industrious virtuoso," for Bacon, conquered the world by direct reading of the book of nature.

Frontispiece, *Sylua syluarum* (1627), in which the story of *New Atlantis* was first published. As the central agent in nature, the sun is identified with God by the inscription of the name of God in Hebrew on its face. Francis Bacon, "New Atlantis," in *Sylua syluarum, or A naturall historie* (1651). Courtesy of Houghton Library, Harvard University.

Pico and Bacon's followers propagated Horace's *Nullius in verba* and called upon readers to trust no book or teacher but to place all faith in practical exploration and firsthand experience. The autodidactic plea spread through the seventeenth century and captured the minds of English natural philosophers, especially those who sided with Parliament in the Civil War and rejected transcendent royal authority. Although he was not of this group, Edward Pococke used the trope of the autodidact to secure his academic career. The stylishly written Arabic manuscript was, in one sense, merely another antiquarian text object, a naive, foreign exploration of certain ideas that had already been discussed by European thinkers who challenged authority by setting tales of autodidacticism and practical natural philosophy on desert islands. To help survive the political turmoil and to secure his job at Oxford, however, Pococke pursued a publishing strategy that served key figures in experimentalism, who in turn gave him patronage and protected his position. In the early 1670s, he finally translated the Arabic manuscript of *Ḥayy Ibn-Yaqẓān* as *Philosophus autodidactus*. In the note to his readers, he pointed out that Ḥayy's autodidactic method may be seen "still as a new method" by the experimentalists.

After two centuries of trends in utopian writing, continuous debates over inductive reasoning, and years of increasing experimentalist writings, Pococke presented his *Philosophus autodidactus* as a medieval predecessor to experimentalism. Ḥayy became an oracle from the past, reinforcing arguments in favor of autodidacticism and the self-directed exploration of nature, while those notions were still being negotiated and before they were solidified in the canon of empiricism represented by Locke's *Essay Concerning Human Understanding*.

Philosophus autodidactus, with its tropes of wilderness, solitude, and autodidacticism, played a role in the construction of arguments in favor of experimentalism, empiricism, and liberal education and gained an enormous following. Translated into various European languages, *Philosophus autodidactus* was at hand for Enlightenment philosophers Locke, Spinoza, Leibniz, Christian Huygens, and many others, who found in it a naive medieval representation of their arguments for the self-directed discovery of nature and the ability of man to shape the best of possible worlds.

Two eminent readers of *Philosophus autodidactus,* Locke and Rousseau, contributed much to its transformation from a program of discovery of nature into an educational program in general. In *Some Thoughts Concerning Education,* Locke drew the contours for liberal education, in which each person can learn and transform himself. In *Emile,* Rousseau presented a program of natural education as the prime agent for social mobilization. John Stuart Mill extended this trend to industrial societies and carried the ideas forward into the future, proposing that poverty and overpopulation could be combated by investing in general education that would transform poor people into civilized and useful citizens.

The liberal stress on education also found its place in children's literature, where the wandering prodigy exploring the world around him became a common narrative in popular culture, in which wild prodigies such as Mowgli and Tarzan came to represent quintessential natural autodidacts and the least alienated beings. In modern times the search for self on a desert island signified a critique of modern alienation. To find one's true self was to surpass the constructed psychic barriers acquired when experiencing modern culture.

Pleas for autodidacticism echoed not only within close philosophical discussions. Struggles over control of knowledge between individuals and establishments were the soil in which such claims grew. The deeply buried and convoluted migration, circulation, and reception of *Hayy Ibn-Yaqzān* illustrate the intricate evolution of social and individual control over knowledge. In twelfth-century Marrakesh, Ibn-Tufayl struggled to tame uncontrolled religious enthusiasm of mystics, subjecting them to philosophical procedures. In fourteenth-century Barcelona, Narbonni called for the liberation of adolescents from communal control, confining them to a self-disciplined life of solitude. In fifteenth-century Florence, Pico della Mirandola sought to cast off the spiritual chains of the heavens by grounding man to the physical reality of nature. Sixteenth-century utopian and scientific societies defied the authority of institutional discipline but called its members to follow restricted, even ascetic, ways of life. And seventeenth-century experimentalists were determined to liberate man from the tutelage of metaphysics and received, constructed categories of knowledge, advocating exclusive reliance on self-discovery of nature.

The various historical nexuses mutually chained one another through the growing process of shifting social control from establishments to the individuals themselves, creating the "modern man" as a solitary, self-controlled individual who utilizes reason to foresee the future outcomes and repercussions of his deeds. The rejection of traditional authorities, thus, did not exclaim chaos, anarchism, or complete freedom. In a mishap, it led to internalization of institutional mechanisms of control. Rather than loss of control, autodidacticism highlighted the need for self-control. It relocated the mechanisms of control from society to the individual, replacing history, tradition, and metaphysics with nature, reason, and natural instincts as the ultimate sources of knowledge.

NOTES

Introduction · The Pursuit of the Natural Self

1. John Locke, *Some Thoughts Concerning Education* (London, 1693), pt. 2, p. 66.

2. John Locke, "Of the Improvement of our Knowledge," in *Essay Concerning Human Understanding* (London, 1690), bk. 4, ch. 12, p. 9.

3. The most popular title of the Tom Telescope series was *The Newtonian System of Philosophy, Adapted to the Capacities of Young Gentlemen and Ladies* (London, 1766). The anonymous writer was James Mitchell (1787–1844), a writer of scientific works, including *The Elements of Natural History* (1819) and *The Elements of Astronomy* (1820).

4. Jean-Jacques Rousseau, *Émile, or On Education*, trans. Barbara Foxley (London, 1911), 1:11, 1:24, 3:564.

5. Sarah Trimmer, *An Easy Introduction to the Knowledge of Nature, and Reading the Holy Scriptures. Adapted to the Capacities of Children* (London, 1799), v–vi.

6. On the influences of Locke and Rousseau on children's literature, see Samuel Pickering, *John Locke and Children's Books in Eighteenth-Century England* (Knoxville: University of Tennessee Press, 1981); Sylvia Patterson, *Rousseau's Émile and Early Children's Literature* (Metuchen, N.J.: Scarecrow Press, 1971); Annette Wannamaker, *Boys in Children's Literature and Popular Culture: Masculinity, Abjection, and the Fictional Child* (New York: Routledge, 2008). Steven Pinker recently subverted the concept of the blank slate in addressing cutting-edge discoveries in linguistics, genetics, and neurology. In a way, the book at hand is the aftermath of his best-selling book *The Blank Slate: The Modern Denial of Human Nature* (New York: Viking, ca. 2002).

7. *The Fables of Bidpai* was translated first into Pahlevi, then into Arabic in the eighth century, Hebrew in the eleventh century, Latin in the twelfth, Spanish in the fourteenth, and Italian in the sixteenth, and by 1570 it had been translated into English. The same book of fables passed through a polytheistic and transmigrational society to a legalistic and monotheistic society where, as the *Kalila wa dimna*, it remains a major educational treatise.

8. Antonio Pastor, *The Idea of Robinson Crusoe* (Watford, U.K., 1930).

9. G. W. Leibniz, *Sämtliche Schriften und Briefe, 1663–1685* (Darmstadt, Ger., 2006), 325, letter 102.
10. See, for instance, Sami S. Hawi, *Islamic Naturalism and Mysticism: A Philosophical Study of Ibn-Tufayl's "Ḥayy bin Yaqṣān"* (Leiden, Neth.: Brill, 1997), 282.
11. Baron Cäy von Brockdorff, "Spinoza's Verhältnis zur Philosophie de ibn Tophail," in *Veröffentlichungen der Hobbes-Gesellschaft* (Kiel, Ger., 1932), 19–31.
12. Voltaire, *Correspondence and Related Documents,* in *The Complete Works of Voltaire*, ed. Theodore Besterman (Geneva, 1968–77), 2:85–135, doc. 310.
13. Scholars of Islamic philosophy looked only at the intertextual relations between *Ḥayy Ibn-Yaqẓān* and other Islamic stories, especially an earlier version of *Ḥayy* written by Avicenna. Scholars of Jewish philosophy have been interested in the ways questions regarding the medieval philosophical concept of the active intellect intersected in *Ḥayy Ibn-Yaqẓān*. See, for instance, Sami Hawi, *Islamic Naturalism and Mysticism: A Philosophic Study of Ibn Ṭufayl's "Ḥayy Ibn-Yaqẓān* (Leiden: Brill, 1974); George F. Hourani, "The Principal Subject of Ibn Tufayl's *Ḥayy Ibn-Yaqẓān*," *Journal of Near Eastern Studies* 15, no. 1: 40; Aaron Hughes, *The Texture of the Divine: Imagination in Medieval Islamic and Jewish Thought* (Bloomington: Indiana University Press, ca. 2004). Such medievalist trends have been disconnected from one other, each offering a philosophical or philological analysis of the text but not a historical investigation of the cultural contexts within which it was written, translated, circulated, and discussed. Additionally, the reception of the story of *Ḥayy Ibn-Yaqẓān* during the Enlightenment has caught the interest of a few scholars wanting to show a sense of continuity between the modern European world and medieval Islamic philosophy. In the 1930s Pastor's *The Idea of Robinson Crusoe* (1930) became the first to engage with the circulation of *Ḥayy Ibn-Yaqẓān,* showing first that the idea of Robinson Crusoe came from Muslim Spain and then that "the Romance of *Ḥayy*" and the concept of life on a desert island came to Muslim Spain from Greek stories about Alexander the Great. Ibn-Tufayl, Pastor shows, transformed that story into a philosophical text about the conjunction of religion and reason. In 1938 Pastor's contemporary, Rudolph Altrocchi, pointed to the relationship between motifs in *Ḥayy Ibn-Yaqẓān* and Dante's *Divine Comedy* and tried to recover historical evidence for the circulation of the Hebrew translation, which, he claims, Dante himself read. Pastor and Altrocchi, however, overlooked the particular historical moments that drew philosophers to translate, circulate, and discuss the story of *Ḥayy Ibn-Yaqẓān*, especially as it factored in humanist Renaissance, anti-astrological discussions and early modern experimentalist controversies. See Antonio Pastor, *The Idea of Robinson Crusoe* (Watford, U.K.: Góngora Press, 1930); Rudolph Altrocchi, "Dante and Tufail," *Italica* 15, no. 3: 125–28. Since the 1930s most of the accounts have fragmented into specific academic areas of study that gave a compressed cultural view of this story. In the late 1990s Lawrence Conrad published *The World of Ibn-Tufayl: Interdisciplinary Perspectives of "Ḥayy Ibn-Yaqẓān*," a transcript of the proceedings of a conference dedicated to the topic. The articles employ literary critique, philological analysis, and comparative philosophy and offer a variety of methodological angles. Whereas most of the discussions focus either on medieval Islamic and Jewish philosophy or on the story's circula-

tion during the Enlightenment, a large gap remains between treatments of the medieval and early modern discussions of the tale. See Lawrence Conrad, *The World of Ibn-Tufayl: Interdisciplinary Perspectives of "Ḥayy Ibn-Yaqẓān"* (Leiden, Neth.: Brill, 1996). More recently, the subtitle of Samar Attar's 2007 book, *Ibn-Tufayl's Influence on Modern Thought*, points to the many writers in the early modern Western tradition who may have drawn upon Ibn-Tufayl. Samar Attar, *The Vital Roots of European Enlightenment: Ibn-Tufayl's Influence on Modern Thought* (Lanham, Md.: 2007). At the other end of the historiographic spectrum, histories of autodidacticism are almost nonexistent, and histories of experimentalism have exclusively focused on European culture, particularly since the late seventeenth century.

Chapter 1 · Taming the Mystic: Marrakesh, 1160s

1. Léon Gauthier counted six extant Arabic manuscripts. One is in the Escorial library but is illegible and damaged; the second, ascribed to Ibn-Tufayl, is considered to be a draft; the third copy is at Dār al-Kuttub in Egypt; a fourth manuscript resides in Algiers. This version served as the source for Gauthier's translation into French. The fifth lies in the British Museum; the sixth, which is the source for this discussion, as well as for Pococke's influential translation, is kept in the Bodleian Library. See Léon Gauthier, *Ibn Thofail, sa vie, ses oeuvres* (Paris, 1909), 43–44.

2. Abū Ḥāmid Muhammad al-Ghazzālī, *Al-Maʿārif al-ʿaqlīyyah wa lubāb al-ḥikmat al-ilahīyya*, Bodleian Library, Oxford University, Pococke 263; ʿAlī Idrīs, ed., *Maʿārif al-ʿaqlīyah wa-lubāb al-ḥikmah al-īlāhīyah* (Sfax, Tunisia: Al-Taʾāḍudīyah al-ʿUmmālīyah lil-Ṭibāʿah wa-al-Nashr, 1988).

3. Most scholarship on *Ḥayy Ibn-Yaqẓān* followed Edward Pococke's view that the treatise essentially represents unaided human reason's ascent to the knowledge of God, ignoring the cultural surroundings, including the controversial reception to al-Ghazzālī's work in al-Andalus. In the early twentieth century, Léon Gauthier proposed the harmony of religion and philosophy as the essential subject of the book. "*Hayy ben Yaqdhān: Roman philosophique d'Ibn Thofail*, Arabic text and French translation by Léon Gauthier, 2nd ed. (Beirut, 1936), ix, xix. Recent scholars have been more concerned with the cultural character presented in the treatise. George Hourani points to the dramatic climax of the work—when Ḥayy enters the cave and gains mystical communion with God through meditation—to argue that the work does not simply reconcile philosophy with religion. Ḥayy's practices, such as his taboos against injury to animal and plant life, Hourani notes, "go far beyond anything in Islam, and have a Pythagorean or Hindu character." George Hourani, "The Principal Subject of Ibn Tufayl's *Hayy Ibn Yaqzan*," *Journal of Near Eastern Studies* 15, no. 1: 40–46, 44; for the interesting connection to Pythagorean and Hindu practices, see Gauthier's *Roman philosophique d'Ibn Thofail*, 81n1. Other meticulous philosophical and philological accounts compare the theories of knowledge in *Ḥayy Ibn-Yaqẓān* with those of Avicenna and Suharawardi. See Yaḥyā ibn Ḥabash Suhrawardī, *Risālāt-i "Ḥayy ibn Yaqẓān*," trans. Ḥamdī Sanadājī and Burhān al-Dīn (Teheran, 1977). They use Ibn-Tufayl's reference to oriental philosophy (*ḥikmat al-Ishrāq*) in the introduction to *Ḥayy Ibn-*

Yaqẓān to argue for the existence of a secretive Avicennian illuminist philosophy. In reaction, Dimitri Gutas forcefully shows that this short mention has been the basis of a centuries-long misunderstanding. In fact, he claims, no secretive Avicennian "oriental philosophy" existed until it was created by Ibn-Tufayl himself. Gutas shows that Avicenna's *ḥikmat al-Ishrāq* differs from the rest of his work only in form and not in substance. Dimitri Gutas, "Ibn-Tufayl on Ibn-Sina's Eastern Philosophy," *Oriens* 34 (1994): 222–41.

4. For a description of the burning, as well as of its reception in the Maghrib, see Kenneth Garden, "Al-Ghazālī's Contested Revival: Iḥyā' 'Ulūm al-Dīn and Its Critics in Khorasan and the Maghrib" (Ph.D. diss., University of Chicago, 2005).

5. Reinhart Dozy, *Histoire des Musulmans d-Espagne* (Leiden, Neth.: Brill, 1932), 3:157–58.

6. See Garden, "Al-Ghazālī's Contested Revival," iv.

7. Al-Ghazzālī, *Iḥyā' 'ulūm al-dīn* [The Revival] (Cairo: Mu'assasat al-Ḥalabī, 1967–68), 1:75–76.

8. The whole program has a theological ramification, since "the *tawḥīd* [oneness of God]," as he said, "stems from the science of unveiling." Al-Ghazzālī, *The Revival*, 4:35.

9. Garden, "Al-Ghazzālī's Contested Revival," 61. Al-Ghazzālī deliberately uses accounts of what the Prophet said, did, or tacitly approved of (Hadith) that were considered weak (*ḍa'īf*) to raise the problem of the trustworthiness and authority of intellectual knowledge, which was a source of much of the criticism that *The Revival* received.

10. Al-Ghazzālī addresses the topic of sincerity as one of the stages through which a Sufi must progress, in bk. 37 of *The Revival*.

11. Quoted in Muḥammad al-Dhahabī, *Siyar a'lām al-nubalā'* (Beirut, 1981), 19:333–34. Italics mine.

12. Abu al-'Abbas Ahmad Ibn Husayn Ibn Qunfudh, *Al-Fārisīyah fī mabādi' al-dawlah al-ḥafṣīyah: Taqdīm wa-taḥqīq Muḥammad al-Shādhilī al-Nayfar wa-'Abd al-Majīd al-Turkī* (Tunis, 1968), 99–100.

13. Ibn-Tufayl, *Ḥayy Ibn-Yaqẓān*, Bodleian Library, Pococke 263, 24b. Italics mine.

14. Ibid., 24b.

15. Ibn-Tufayl's "fa-kāna mā kāna mimmā lastu adhkuruhu / fa-ẓunna khayran wa-lā tas'al 'ani l-khabari" could also mean "and it was what it was, of which I do not speak [or "which I do not recall"]; thus, think the best and do not ask me about the matter." Ibn-Tufayl, *Ḥayy ibn Yaqẓān*, 24b.

16. Ibn-Tufayl, *Ḥayy ibn Yaqẓān*, 56a.

17. Ibid., 56b.

18. Ibid., 56b-57a.

19. Ibid., 57b-58a.

20. By the mid-twelfth century we find Sufis such as Ibn Barrajān, Ibn al-'Arīf, and Ibn Qasī writing mystical treatises and the *murīdūn*, a Sufi order led by Abū al-Qāsim Ibn Qasī, mounting a revolt against the Almoravids. See accounts by Maribel

Fierro, "La religion," in *Historia de España*, vol. 8, *Los reinos de taifas: Al-Andalus en el siglo XI*, ed. Ramón Menéndez Pidal (Madrid, 1994), 437; "The Polemic about the Karāmāt al-Awliyā' and the Development of Ṣufism in al-Andalus (Fourth/Tenth–Fifth/Eleventh Centuries)," *Bulletin of the School of Oriental and African Studies* 55 (1992): 236–49; Vincent Cornell, *Realm of the Saint: Power and Authority in Moroccan Sufism* (Austin, Tex., 1998).

21. Rachid el Hour, "The Andalusian Qaḍī in the Almoravid Period: Political and Judicial Authority," *Studia Islamica* 90 (2000) 67–83. See Garden, "Al-Ghazālī's Contested Revival," 144–55.

22. Fernando Rodríguez Mediano, "Biografias Almohades en el Taṣawwuf de al-Tālidī," *Estudios onomástico-biográficos de al-Andalus* 10 (2000): 167–93, 177.

23. Garden, "Al-Ghazālī's Contested Revival," 9.

24. The description of the second burning is taken from Garden, "Al-Ghazālī's Contested Revival," and Cornell, *Realm of the Saint*, 21–22.

25. Ḥasan Ibn ʿAlī Ibn al-Qaṭṭān al-Marrākushī, *Nuẓum al-jumān li-tartīb mā salafa min akhbār al-zamān* (Beirut, 1990), 70–73.

26. Madeleine Fletcher argues that Ibn-Tūmart's teaching assigns a radically important role to reason in theology, declaring reason to be a source of religious doctrine along with the scriptures (the Quran and the Hadith). Madeleine Fletcher, "The Almohad Tawḥīd: Theology Which Relies on Logic," *Numen* 38, no. 1 (1991): 110–27.

27. Muḥammad Ibn-Tūmart, *Aʿazz mā yuṭlab wa afḍal mā yuktasab wa anfas mā yudhkar wa aḥsan mā yuʾmal: Al-ʿilm alladhī jaʿalahu Allāh sabab al-hidāya ilā kulli khayr*. The text is contained in Ignaz Goldziher, *Le livre de Mohammed Ibn Toumert mahdi des Almohades* (Algiers: P. Fontana, 1903), 30, citation from Cornell, *Realm of the Saint*, 93.

28. Vincent Cornell, "Understanding Is the Mother of Ability: Responsibility and Action in the Doctrine of Ibn Tūmart," *Studia Islamica*, no. 66. (1987): 71–103.

29. Muḥammad Ibn-Tūmart, *Aʿazz mā yuṭlab wa afḍal mā yuktasab wa anfas mā yudhakhar wa aḥsan mā yuʾmal*, 30, citation from Cornell, *Realm of the Saint*, 96.

30. In Ibn-Tūmart's view, the followers of his doctrine impart guidance to others in three ways: by indication (*al-ishāra*), through the written word (*al-kitāba*), and by way of parables and examples (*al-ʿibāra*). Muḥammad Ibn-Tūmart, *Aʿzz mā yuṭlab wa afḍal mā yuktasab wa anfas mā yudhakhar wa aḥsan mā yuʾmal*, 33, citation from Cornell, *Realm of the Saint*, 99.

31. I. S. Allouche, *Chronique anonyme* (Paris: Institut des Hautes Étude Marocaines, 1936), 89–90. A. Huici Miranda, trans., *Al-hulal al-Mawshiyya* (Tetuan, Morocco, 1951), 131, citation from Fletcher, "The Almohad Tawḥīd," 112.

32. The efficiency of the bureaucracy was traditionally attributed to the strict hierarchical sociopolitical structure that played a central role in solidifying and implementing its political theology. See J. F. P. Hopkins, "The Almohade Hierarchy," *Bulletin of the School of Oriental Studies* 16, no. 1. (1954): 93–112.

33. Reinhart Dozy, ed. *The History of the Almohades, by ʿAbdu al-Wāhid al-Marrākushī* (Leiden, Neth.: Brill, 1881), 172–74.

34. Ibid.

35. Ibn Rushd al-Jadd (d. 1126) and the theologian Muhammad b. Khalf b. Mūsā al-Anṣārī al-Ilbīrī (d. 1142–43) corresponded regarding *The Revival* and Ibn-Ḥamdīn's refutation of it. Ibn Rushd's rational objection to al-Ghazzālī, as well as his qualifications of the arguments of those who opposed it, was transmitted to his grandson, Averroes, who was first introduced to the Almohad court by Ibn-Tufayl himself. Al-Ilbīrī wrote a refutation of al-Ghazzālī's treatment of the soul. Ibn Rushd found many gross mistakes in this refutation and called on him to correct those mistakes. As a result, he seemed to be defending al-Ghazzālī. See Abū al-Walīd Ibn Rushd al-Jadd, *Masā'il Abī al-Walīd Ibn Rushd*, 1:346; cited in Garden, "Al-Ghazālī's Contested Revival," 186.

36. For the life of Ibn-Tufayl, see Gauthier, *Ibn Tufayl, sa vie, ses oeuvres*. The reports quoted can be found in Dozy, *History of the Almohades*, 174.

37. Maribel Fierro, "Opposition to Sufism in al-Andalus," in *Islamic Mysticism Contested: Thirteenth Centuries of Controversies and Polemics*, ed. Frederick de Jong and Bernd Radtke (Leiden, Neth.: Brill, 1999), 174–97, 190.

38. Ibid., 196.

39. Ibn Bashkuwāl (d. 1183), a jurist intellectually active during Ibn-Tufayl's time, tried to give jurists new forms of religious authority that had less to do with theology and jurisprudence than with the ability to manipulate divine forces and popularize practices usually associated with Sufis. During the 1160s and 1170s the portrayal of jurists with saintly attributes spread widely. Ibid., 190.

40. Avempace was born in Saragossa toward the end of the eleventh century. He was educated at Fez and moved to Seville in 1118, where he lived until his death in 1138, two decades before Ibn-Tufayl set out to write his novel. Ibn-Tufayl praised his wisdom and learning highly in *Ḥayy Ibn-Yaqẓān* but noted that his premature death prevented him from publishing any major works, leaving his more important writings incomplete and little known. One of Avempace's methodological writings that survived only in its Hebrew translation was *Regime of Solitude*, which praises the solitary life of philosophers, a point echoed in *Ḥayy Ibn-Yaqẓān*. Avempace did not accept the Aristotelian principle that the velocity of an object is in inverse ratio to the resistance of the medium and, therefore, a motion in void would be infinite, and so impossible. In contrast to this traditional view, Avempace believed in the possibility of motion in a void. For him, the medium was not essential to natural motion at finite velocity, as Aristotle held, because the speed was determined by the difference, rather than the ratio, between the densities of body and medium. Averroes' great commentary on Aristotle's *Physics*, especially on text 71 of book 4, was the point of departure for scholastic commentary on the *Physics* from the time of Albertus Magnus to that of Galileo. Salomon Munk, *Mélange de philosophie juive et arabe* (Paris, 1857), 383–410. Munk gives an analysis of Avempace's *Régime du soitaire* based on a Hebrew summary of the work; this suggests that *Ḥayy Ibn-Yaqẓān* represented Ibn-Tufayl's interpretation of his teacher's teachings. The Arabic text of this work was published with a Spanish translation by Miguel Asín Palacios, *Avempance: "El régimen del solitario,"* ed. and trans. Miguel Asín Palacios (Madrid, 1948).

41. Ibn-Tufaly, *Ḥayy Ibn-Yaqẓān*, 29a–29b.
42. Al-Mas'ūdī, *Murūj al-dhahab wa-ma'ādain al-jawhar* (Beirut, 1965), 2:112.
43. Ibn-Tufayl, *Ḥayy Ibn-Yaqẓān*, 29b.
44. See Berthold Laufer, "Asbestos and Salamander, an Essay in Chinese and Hellenistic Folk-Lore," *T'oung Pao*, 2nd ser., 16, no. 3 (1915): 299–373.
45. Buzurg ibn Shahriyār, *'Ajāib al-Hind: Barruhā wa-baḥruhā wa-jazā'iruhā/li-Buzurk ibn Shahriyār al-Nākhidhāh al-Ramhurmurzī*; Taḥqīq 'Abd Allāh Muḥammad al-Ḥibshī Abū Ẓaby (United Arab Emirates: al-Mujamma' al-Thaqāfī, 2000); *Livre des merveilles de l'Inde, par Bozorg Fils de Chahriyâr de Râmhormoz; Texte Arabe, publié d'après le manuscrit de M. Schefer . . .*, trans. Marcel Devic (Leiden, Neth.: Brill, 1883–86); *The Book of the Marvels of India*, trans. from the French by Peter Quennell (London: Routledge, 1928).
46. Buzurg ibn Shahriyār, *'Ajāib al-Hind*, 72.
47. On long-distance trade from the Indian Ocean, see K. N. Chadhuri, *Asia Before Europe: Economy and Civilization of the Indian Ocean from the Rise of Islam to 1750* (New York: Cambridge University Press, 1990), 88–91.
48. See Andre Wink, *Al-Hind, the Making of the Indo-Islamic World*, vol. 1, *Early Medieval India and the Expansion of Islam* (Leiden, Neth.: Brill, 2002), 65–109.
49. See James S. Romm, *The Edges of the Earth in Ancient Thought: Geography, Exploration, and Fiction* (Princeton, N.J.: Princeton University Press, 1992).
50. For the classic perception of Taprobane, see D. P. M. Weerakkody, *Taprobanê: Ancient Sri Lanka as Known to Greeks and Romans* (Turnhout, Belg., 1997).
51. Pliny, *HN*, VI 81–91, VII 2.30.
52. S. D. Goitein, "From the Mediterranean to India: Documents on the Trade to India, South Arabia, and East Africa from the Eleventh and Twelfth Centuries," *Speculum: A Journal of Medieval Studies* 29, no. 2, pt. 1 (1954): 181–97.
53. Ibid., 186.
54. Buzurg ibn Shahriyār, *'Ajāib al-Hind*, 39–41.
55. 'Alī ibn Mūsá Ibn Sa'īd, *Kitāb al-jughrāfiyā* (Beirut, 1970), 89.
56. See Fedwa Malti-Douglas, *Woman's Body, Woman's Word: Gender and Discourse in Arabo-Islamic Writing* (Princeton, N.J.: Princeton University Press, 1991), ch. 4. Ibn-Tufayl's description of the wondrous tree had some other later cultural ramifications. From the fourteenth century onward, the symbol of the Wāqwāq's tree that bears women as fruits has been incorporated into cultural objects by imprinting its image on carpets and miniatures. See Gertrude Robinson, "An Unknown Sixteenth-Century Persian Carpet," *Burlington Magazine for Connoisseurs* 72, no. 420 (1938): 102–5.
57. Henri Breuil, *Rock Paintings of Southern Andalusia: A Description of a Neolithic and Copper Age Art Group* (Oxford: Clarendon, 1929).
58. See, for instance, the work of David Lewis-Williams, *Believing and Seeing: Symbolic Meanings in Southern San Rock Paintings* (New York: Academic Press, 1981).
59. Claude Lévi-Strauss, *The Savage Mind* (Chicago: University of Chicago Press, 1968), 1–35.
60. 'Alī ibn Abī al-Ḥazm Ibn al-Nafīs, *The Theologus Autodidactus of Ibn al-Nafīs*,

edited with an introduction, translation, and notes by Max Meyerhof and Joseph Schacht. (Oxford: Clarendon Press, 1968).

Chapter 2 · Climbing the Ladder of Philosophy: Barcelona, 1348

1. The anonymous Hebrew translation of *Ḥayy Ibn-Yaqẓān* had circulated among Jewish philosophical circles before Narbonni came along. Maimonides himself was acquainted with the story and made implicit mention of it in his *Guide for the Perplexed*. Moses Maimonides, *The Guide for the Perplexed*, trans. Michael Shwarz, 2 vols. (Tel-Aviv, 2002), vol. 2, pt. 2, ch. 17. This was, perhaps, another incentive for his followers to look for the original. The introduction of Maimonides' works in Spain and Provence-Catalonia generated various controversies. In 1202 Rabbi Meir HaLevi Abul'afia of Toledo accused Maimonides of not following or believing in the principle of resurrection and expressed a general dissatisfaction with his equivocal treatment of other theological principles. Maimonides' followers dismissed these accusations, arguing that they represented an incompetent reading of the work. In 1203 Sheshet HaNasi of Narbonne sent a letter to the elders of the town of Lunel that attacked Abul'afia's narrow understanding of the questions surrounding resurrection. The ladder of philosophy, he wrote, offers the possibility not only of moving from life to death and then on to the otherworld but also of achieving, in life before death, union with the otherworld. "I saw in a book of one of the greatest philosophers that a soul that would aspire to return to its creator . . . should not do a thing other than love him and should refrain from eating impure things or even tolerat[ing] corporeal impurities . . . and only then will God take the soul and make it one of His angels." Signing his letter as "Sarkastic Yehi bi 'Am" [sarcastic, a people will live in me], HaNasi echoes the the title *Ḥayy Ibn-Yaqẓān*. The Hebrew text of this letter was first published by Alexander Marx in the *Jewish Quarterly Review* 35 (1944): 414–28, citation at 422.

2. In his introduction to *Ḥayy Ibn-Yaqẓān*, Pococke mentioned that he also relied on extant copies of the Hebrew translation of Narbonni, and there are other indications that copies of this text circulated among Jewish scholars in the Mediterranean. See Edward Pococke, *Philosophus autodidactus* (Oxford, 1671), "Ad lectorem," p. A.

3. Noel Coulet, "Les Juifs en Provence au Bas Moyen Age: Les limites d'une marginalite," in *Minorités et marginaux en France méridionale et dans la péninsule ibérique (VIIe–XVIIIe siècles): Actes du colloque de Pau, 27–29 mai 1984* (Paris: Editions du Centre national de la recherche scientifique, 1986), 203–6.

4. Even when the community was integrated into the Christian society, it still regulated its internal affairs and, as in Perpignan, possessed "its assemblies, statues, and its own magistrates." Pierre Vidal, "Les Juifs des anciens comtés de Roussillon et de Cerdagne," *Revue des études juives* 15 (1887): 19–33. Recent scholarship has revised this tradition and argues that the Crown never gave up its right to intervene in the life of the Jewish community, and only in certain matters could Jews maintain an independent life. Maurice Kriegel, for instance, has argued that the degree of Jewish submission to and dependence on royal or seigniorial authority precludes any notions of a medieval Jewish state within a state. Maurice Kriegel, *Les Juifs à la fin du Moyen*

Age dans l'Europe méditerranéenne (Paris, 1979), 113–14. Instead of belonging to the *universitas*—a royal designation of juridical identity—the community in Perpignan held a special status of *aljama judeorum*, a corporate body with its own identity and legal standing. The name *aljama*, from the Arabic *al-Jama'a* (a group or community), derived from the status given to a Muslim community of Perpignan, as in other towns in Christian Spain, a community that ceased to exist in the second half of the thirteenth century. Philip Daileader, *True Citizens: Violence, Memory, and Identity in the Medieval Community of Perpignan, 1162–1397* (Leiden, Neth.: Brill, 2000).

5. Most scholars who have paid any attention to Narbonni (born in the late thirteenth century, died ca. 1362) have tended to view him as a thinker who highlighted the nuances of contemporary intellectual issues and have assumed that his choices and themes were dictated by strict intellectual preferences. Some saw him as radicalizing Maimonides by incorporating Averroeist views and emphasizing God as the "necessary being" rather than as a transcendent supernatural being. Few scholars have looked at the relationship between his personal experience and the intellectual choices that made him identify with Ḥayy Ibn-Yaqẓān. Just as Ḥayy lived in solitude on his island, Narbonni distanced himself from his community of Perpignan, a Mediterranean town that bordered southern France and Spain. Maurice Hayoun, "Moshe Narbonni: Be'ayot haNefesh u-Kohoteha" [Moses Narbonni: Problems of the soul and its powers], *Da'at* 23 (1989): 65–88. E. I. Rosenthal, "The Place of Politics in the Philosophy of Ibn Bajja," *Islamic Culture* 25 (1951): 187–211; "Political Ideas in Moshe Narbonni's Commentary on Ibn Tufail's Hay B. Yaqzan," in *Hommage à Georges Vajda*, ed. G. Nahon and C. Touati (Leuven, Belg., 1980), 227–34. See also Georges Vajda, "Comment le philosophe Juif Moise de Narbonne, commentateur d'Ibn Tufayl, comprenait-il les paroles extantiques (satahāt) de Soufis?" in *Mélanges Georges Vajda: Études de pensée, de philosophie et de littérature juives et arabes; In memoriam*, ed. G. E. Weil (Hildesheim, Ger.: Gerstenberg, 1982), 275–81. On aligning Narbonni with radical Aristotelianism, see Colette Sirat, "Perkei Moshe le-Moshe Narbonni" [Moshe's chapters of Moshe Narbonni], *Tarbiz: A Quarterly for Jewish Studies* 39 no. 3 (1970): 287–306. A few scholars have glancingly touched on that relationship. Larry Miller, for instance, argues that the biographical insertions in the introduction of *Yehiel Ben-'Urie* indicate that Narbonni personally identified with Ḥayy Ibn-Yaqẓān. Larry Miller, "Philosophical Autobiography: Moshe Narbonni's Introduction to his Commentary on Ḥayy Ibn-Yaqẓān," in *The World of Ibn Tufayl*, ed. Lawrence Conrad (Leiden, Neth.: Brill, 1996), 232. In the same vein, Gitit Holtzman argues that Narbonni's tragic life led him to follow Averroes' theory of the soul, suggesting that perhaps Narbonni found solace in the knowledge that his inner light could lead him to communion with God and to ultimate felicity. Gitit Holzman, *Torat ha-nefesh veha-sekhel be-haguto shel R. Mosheh Narbonni: Al pi beurav le-khitve Ibn Rushd, Ibn Tufil, Ibn Bag'ah ve-Algazali* (Ph.D. diss., ha-Universitah ha-Ivrit, Jerusalem, 1996), 26.

6. Narbonni mostly wrote commentaries on Hebrew translations of contemporary philosophy, in which he extensively engaged with the philosophical questions of the soul's immortality, the fundamental laws of nature, and the possibility of union with the active intellect. These include works by Averroes and al-Ghazzālī. Later in

his life he wrote commentaries on *Ḥayy Ibn-Yaqẓān* and Avempace's *Regime of the Solitude*. Finally, before he died, his philosophical interest culminated in his most influential work, a commentary on Maimonides' *The Guide for the Perplexed*. His didactic attitude, so evident in his commentary on *Ḥayy Ibn-Yaqẓān*, is clear in the only personal philosophical essay he wrote, *A Letter on the Perfection of the Soul* (Igeret shelemut hanefesh). Not written in the form of commentary, the letter is addressed to Narbonni's son with the express purpose of giving him a philosophical education. Alfred Lyon Ivry, "Moses of Narbonne's Treatise: The Perfection of the Soul; A Partial Edition from the Paris MS" (Ph.D. diss., Brandeis University, 1963).

7. Narbonni also mentioned that he used to discuss these issues with another teacher-colleague, Joseph Ibn Wakar, a philosopher and famous Kabbalist. See Narbonni's commentary on *The Guide for the Perplexed*, in *Sheloshah ḳadmone mefarshe ha-Moreh* (Jerusalem, 1960), pt. 1, ch. 28. On Joseph Ibn Wakar, see Gershom Sholem, "Sifro HaAravi shel R' Yoseph N' Vakar 'al HaKabala ve HaFilosophia," *Kiryat Sefer* 20 (1970): 153–62. In his writing on medicine, Narbonni mentioned *Orah Hayyim* and a physician called Avraham Kashlari. See Harry Friedenwald, *The Jews in Medicine* (New York, 1967), 2:682–83.

8. See E. Renan, *Les écrivains Juifs français du XIVe siecle* (Paris, 1983), 320n1; see also Isidore Loeb, "Liste nominative des Juifs de Barcelone," *Revue des études juives* 4 (1882): 57–77, 69. Pierre Vidal, a late nineteenth-century historian of Perpignan, mentions Bonjuhes Bellashom, a Jew who lived in Perpignan in the early fifteenth century and who had an extensive library. Pierre Vidal, "Les Juifs des anciens comtés de Roussilon et de Cerdagne," *Revue des études juives* 16 (1888): 181.

9. See Henry Gross, *Galia Judaica* (Paris, 1897), 401–3. It was the home of the Jewish aristocracy of Provence. On the place of the president (*nasi*) from Narbonne in shaping the social and intellectual life of the Jews of Provence, see Shlomo H. Pick, "The Jewish Communities of Provence before the Expulsion in 1306" (Ph.D. diss., Bar-Ilan University, 1996), 112–61. Provence was also a center of Jewish-Christian controversies in the thirteenth century. Siegfried Stein, "Jewish-Christian Disputations in Thirteenth-Century Narbonne" (inaugural lecture delivered at University College, London, October 22, 1964) (London: H. K. Lewis, 1969).

10. *The epistle on the possibility of conjunction with the active intellect, by Ibn Rushd; with the commentary of Moses Narbonni; a critical edition and annotated translation by Kalman P. Bland* (New York: Jewish Theological Seminary of America, 1982), 149–50.

11. See Holzman, *Torat ha-nefesh veha-sekhel be-haguto shel R. Mosheh Narbonni*, 9.

12. As Narbonni reports in his commentary on *Regime of Solitude*, "After we finished the commentary on *Ḥayy Ibn-Yaqẓān*, I thought I should also write on Avempace's *Regime of Solitude*. After I obtained this manuscript and while writing, I fled to Valencia." *Hanhagat Hamitboded*, Bayerische Staatsbibliothek, Munich, cod. Hebr. 59, 140a. The following years Narbonni moved from Valencia to Toledo, and in 1362 he completed his magnum opus, a commentary on *The Guide for the Perplexed*, in Suria, on the northern Iberian Peninsula. At the end of the commentary he wrote, "It was completed here in Suria, in April 1362, as I am preparing to return now to

my homeland and hometown Perpignan." Moshe Narbonni, *Biur Moreh Nevochim*, Bibliothèque Nationale, Paris, Heb. 696, ms. 70.

13. Narbonni writes that the people of Perpignan "choose me and I choose them, to influence them [to] the truth as it was revealed to me, and they stimulated me to clarify the meaning of this epistle." Moshe Narbonni, *Yehiel Ben-'Urie*, Bayerische Staatsbibliothek, Munich, cod. Hebr. 59, 2b.

14. Aba Mari, Sefer minhat kanaut, vol. 5 of *Teshuvot ha-Rashba: Le-Rabenu Shelomoh Aderet: Teshuvot ha-shayakhot la-Mikra midrash ye-de'ot ye-tsuraf la-hen Sefer Minhat kena'out le-R[aba] Mari de-Lonil*, by Solomon ben Abraham Adret (Jerusalem: Mosad ha-Rav Kuk, 1990), letter 51.

15. An indication of how central Arabic sources were to Provençal-Catalonian Jewish intellectuals can be seen in the most influential translator, Yehuda Ibn-Tibbon (1120–90), who wrote in his will that "you should know that the greatest scholars of our people did not have greatness and intellectual virtues, but only through writings in Arabic." *Tsava'ot ge'one Yisrael*, ed. Abrahams Israel, 1:55–85 (Philadelphia: Ha-Hevrah ha-Yehudit le-hotsaat sefarim asher ba-Amerika, 1926).

16. "Igeret Le-Rabbi Shemuel Ibn-Tibon be-'Inyane tirgum HaMoreh," in *Igrot HaRambam* [Letters of Maimonides] (Jerusalem: Shilat, 1986), pt. 2, p. 550.

17. Moses Maimonides, *The Guide for the Perplexed*, pt. 3, ch. 51.

18. See the eclectic commentary on Mss. Parma, De Rossi [collection], 272, 78a, as presented by James T. Robinson, "Samuel Ibn Tibbon's Commentary on Ecclesiastes" (Ph.D. thesis, Harvard University, 2002), 188.

19. See, for instance, Ibn-Tibbon's commentary on Aristotle's *Meteorology*, *Otot ha-shamayim: Samuel Ibn Tibbon's Hebrew Version of Aristotle's Meteorology*, a critical edition with introduction, translation, and index by Resianne Fontaine (Leiden, Neth.: Brill, 1995).

20. Mari, *Sefer minhat kanaut*, letter 1, pp. 13–14.

21. Solomon ben Abraham Adret, *Teshuvot ha-Rashba: Le-Rabenu Shelomoh Aderet; Teshuvot ha-shayakhot la-Mikra midrash ye-de'ot ye-tsuraf la-hen Sefer Minhat kena'out le-R[aba] Mari de-Lonil* (Jerusalem: Mosad ha-Rav Kuk, 1990), vol. 1, sign 415, pp. 225–26.

22. Ibid., vol. 1, sign 416, p. 227.

23. Mari, *Sefer minhat kanaut*, letter 3, p. 16; letter 5, p. 24; letter 12, p. 39; letter 13, p. 39.

24. Adret, *Teshuvot ha-Rashba*, vol. 1, sign 416, p. 227.

25. Mari, *Sefer minhat kanaut*, letter 24, p. 57.

26. Moshe Halbertal, *Ben Torah Le-Hochma: Rabbi Menachem Hamairi u Ba'ali Hahalach Ha-Maimonim be Provence* [Between Torah and wisdom: Rabbi Menachem ha-Meiri and the Maimonidian Halakhists in Provence] (Jerusalem: Hebrew University Magnes Press, 2000), 152–80.

27. Rabbi Shimon Ben-Yosef, *Hoshen Mishpat, Tif'eret sevah: divre sofrim*. . . . (Berlin: L. Gershel, 1884), 165–66.

28. Mari, *Sefer minhat kanaut*, letter 101, p. 162.

29. Ibid., letter 10, p. 36; letter 11, p. 36; letter 12, p. 37.

30. See Richard Wilder Emery, *The Jews of Perpignan in the Thirteenth Century: An Economic Study Based on Notarial Records* (New York: Columbia University Press, 1959), 41n4.

31. For instance, Aderet and his supporters searched for other active members of their circle and ultimately figured out the identity of the person who gave the original speech that sparked the controversy, "the philosophizing Rabbi Levi." They urged Shmuel Ha-Salmi, who hosted him in his house, "to expel him and disconnect any social relationship with him." Mari, *Sefer minhat kanaut*, letter 17, p. 41.

32. Moses Narbonne, *Beur le-sefer moreh navochim* (Vienna: Aus der K. K. Hof und Staatsdruckerei, 1832), republished in *Sheloshah kadmone mefarshe ha-moreh* (Jerusalem, 1960), pt. 1, ch. 50.

33. Ibid.

34. Ibid., pt. 2, ch. 19.

35. *Mamar be behira*, published and annotated by Morice Hayoun, in "Lépitre du libre arbitare de Moise de Narbonne," *Revue des études juives* 141 (1982): 139–67. For the polemics, see Carlos del Valle, "El libro de las batallas de Dios, de Abner de Burgos," in *Polémica judeo-cristiana: Estudios*, ed. Johann Maier and Carlos del Valle Rodríguez (Madrid: Aben Ezra Ediciones, 1992).

36. Narbonni, *Yehiel Ben-'Uriel*, 4a.

37. Ibid., 30b, 31b.

38. In mentioning Avempace's *Regime of Solitude*, Narbonni pointed to the quintessential medieval work on the solitary contemplative life. And to drive home the connection, he attached a summary of Avempace's *Regime of Solitude* (Tadbīr al-mutaūḥid) to the end of *Yehiel Ben-'Uriel* with the claim that it served as Ibn-Tufayl's source. That was a deliberate reference, since he already planned to compose a commentary on the *Regime of Solitude* for his next project, which he had already begun in 1349 in Cervera, right after completing *Yehiel Ben-'Uriel*. Narbonni, *Yehiel Ben-'Uriel*, 122b.

39. Ibid., 166a.

40. Ibid., 3b.

41. Ibid., 51b.

42. Ibid., 46a. In his citations and references to Maimonides' work, Narbonni echoes a disagreement between Maimonides and Nahmanides regarding the proper way to study. In the introduction to the third part of *Guide for the Perplexed*, Maimonides mentions that according to the Jewish sages the best way to understand the secrets of the Torah, including metaphysics and cosmogony, is to receive the teachings from a teacher. Since the chain of command was broken, however, metaphysics should be unserstood through one's own intellectual ability.

43. Narbonni, *Yehiel Ben-'Uriel*, 9b.

44. Ibid., 110b. The inspiration Narbonni drew from a mystical writer like al-Ghazzālī implies a connection between autodidacticism, sudden perception, and ecstatic Kabbalah. Moshe Idel argues that Narbonni was among the first philosophers also associated with Kabbalist ideas, especially the ecstatic Kabbalah of Avraham

Abulaifia, which included a description of spiritual techniques for mystical union that come as sudden perception of the secrets of the cosmos. Perhaps not coincidentally, Abulafia was also banished by Aderet. Moshe Idel, *Studies in Ecstatic Kabbalah* (New York: State University of New York Press, 1988), 63–73.

45. Narbonni, *Yehiel Ben-'Uriel*, 2b.

46. Ibid., 114b. Although according to biblical tradition Moses, Aaron, and Miriam died of mitat neshika, Narbonni is probably relying on the meaning of this phrase given by Maimonides, who writes in *The Guide for the Perplexed* (pt. 3, chap. 51) that "as long as his body strengths will weaken and the light of passion will turn off, the power of his intellect will strengthen and in this great pleasure and joy he will go on until the soul will be separated from the body." See *Guide for the Perplexed*, pt. 3, ch. 51. Narbonni connected notions of solitary life, autodidacticism, and sudden perception to the constant autobiographical complaint that he faced death wherever he went. He describes the distraction of the Jewish community in Cervera in 1349 and the martyrdom of young and old Jews. Narbonni says that their choice of martyrdom represented an opportunity to gain communion with God and to achieve ultimate felicity. "I witnessed the persecution and extension of many communities of our people," he sorrowfully relates, "and the wise men gave themselves, young and old, to martyrdom, and chose death over life in order to take the true life." Moses Narbonni, *Sefer Haderushim haTiv'iyot ve-haEloiyot le-Ibn-Rushd*, Bibliothèque Nationale, Paris, Heb. 988, ms. 66b.

47. Mari, *Sefer minhat kanaut*, Ben-Yehiel letters 51, 52.

48. Narbonni, *Yehiel Ben-'Uriel*, 166a.

49. Harry Wolfson, *Crescas' critique of Aristotle: Problems of Aristotle's Physics in Jewish and Arabic philosophy* (Cambridge, Mass.: Harvard University Press, 1929).

50. Joseph Delmedigo, *Sefer melo hofnayim*, ed. Avraham Gaiger (Berlin, 1860), 18.

51. See the opening poem in which Delmedigo is described as a "'wandering boy'"; see also his description of a public dispute on mathematics in which Delmedigo describes himself as a "young boy." Joseph Delmedigo, *Sefer elim* (Amsterdam, 1629), 176–77.

52. See Muhammad Ibn Tufayl, *Ḥayy Ibn-Yaqẓān*, Bayerische Staatsbibliothek, Munich, cod. Hebr. 59.

Chapter 3 · Defying Authority, Denying Predestination, and Conquering Nature: Florence 1493

1. Giovanni Pico della Mirandola, *Hayy Ibn Yaqzan*, Biblioteca Universitaria di Genova, cod. A, IX, ms. 29, fols. 79v–116r.

2. Recently, a team of scientists exhumed Pico's and Poliziano's corpses and subjected them to a battery of tests. In February 2008 they concluded that both men had been poisoned, Poliziano in September and Pico in November, 1494. *London Daily Telegraph*, February 7th, 2008.

3. Savonarola frequently preached that the destiny of man is already inscribed by

the grace of God. See Girolamo Savonarola, *Prediche sopra Amos e Zaccaria*, ed. Paolo Ghiglieri (Rome, 1971), 1:3–4.

4. Leonardo Nogarola, a clergyman in Ferrara, is known primarily as the brother of Isotta Nogarola, who composed the famous *Dialogue on Adam and Eve*, in which she discussed the relative sinfulness of Adam and Eve—thereby opening up a centuries-long debate in Europe on gender and the nature of woman. Gianfranco Fioravanti, "Pico e l'Ambiente Ferrarese," in *Giovanni Pico della Mirandola: Convegno internazionale di studi nel cinquecentesimo anniversario della morte*, ed. Gian Carlo Garfagnini (Florence, 1997), 159.

5. Tito Vespasiano Strozzi, "Ad eundem Johannem Picum Mirandulanum, quod perditis belli temporibus, ut amores suos caneret rogavisset," in *Strozii poetae pater et filius*, ed. Aldo Manuzio (Venice, 1515), 215v. The poem is printed in Giovanni Di Napoli, *Giovanni Pico della Mirandola e la problematica dottrinale del suo tempo* (Rome, 1965), 37.

6. Giovanni Pico della Mirandola, *Oratio de hominis dignitate*. (Università degli Studi di Bologna and Brown University, 1999), 26.

7. Ibid., 28.

8. Cited from John Addington Symonds, *Renaissance Italy* (London, 1877), 329.

9. "Lucio Phosphorus, Bishop of Segni, to his dear Angelo Poliziano," in *Angelo Poliziano: Letters*, ed. and trans. Shane Butler (Cambridge, Mass.: I Tatti Renaissance Library, Harvard University Press, 2006), 181, letter 3:14.

10. Niccolò Machiavelli, *History of Florence and of the Affairs of Italy from the Earliest Times to the Death of Lorenzo the Magnificent*, trans. Felix Gilbert (New York, 1960), bk. 8, ch. 7.

11. Early on, historians of the Renaissance depicted Pico's work as marking the historical shift between medieval obedience to nature and the early modern aspiration to control nature. Jacob Burckhardt (1860), for one, went so far as to argue that as a result of the publication of Pico's *On the Dignity of Man* and later the *Disputations*, "differences of birth . . . lost their significance in Italy," to the extent that "astrologists ceased to publish their doctrines." Jacob Burckhardt, *The Civilization of the Renaissance in Italy* (New York, 1958), 2:351, 492. In his influential account of the Italian Renaissance, *The Individual and the Cosmos* (1927), Ernst Cassirer asserts that Pico's *Oration on the Dignity of Man* and *Disputationes* shifted medieval natural philosophy into a new mathematico-physical theory of nature. "With one blow," Cassirer famously puts it, "[Pico] destroys the sphere of the influence of astrology." Cassirer proposes that fundamental epistemic objects of proximate principles (*proxima principia*) and true cause (*vera causa*) led to the shift. Ernst Cassirer, *The Individual and the Cosmos in Renaissance Philosophy* (New York, 1963), 115. Later, that image of Pico as the originator of a radical historical break became blurred. In her *Giordano Bruno and the Hermetic Tradition*, Frances Yates transforms Pico from "a modern liberal" thinker into a hermetic Renaissance magus. She reconciles the seeming contradiction between Pico's embrace of natural magic and Kabbalah with his vehement rejection of astrology by stressing natural magic as the most ancient universal wisdom about nature, serving as the hermetic phase that led to the scientific revolution. Accordingly,

Pico represented a magus refashioning himself into a protoscientist, armed with free will, dignity, resourcefulness, and the curiosity to engage directly with nature and control his destiny through science. Frances Yates, *Giordano Bruno and the Hermetic Tradition* (London, 1964), 116. More recently, Anthony Grafton portrays Pico as a highly skilled reader of texts. In his essay "Giovanni Pico della Mirandola: Trials and Triumphs of an Omnivore," Grafton argues that Pico read intensively, positing reading as the ultimate vocation of the intellectual. Through his rigorous reading, Pico sought to collect and synthesize the pieces of the vital universal truth written by past and present scholars into one philosophical system, even while subverting textual authority. For Grafton, Pico can be best understood not as an originator of modern science but rather as the person who first historicized astrology. Thus in writing the *Disputationes* Pico foreswore his former admiration of ancient oriental wisdom and replaced it with contempt for the Eastern nest of astrology: "The credulous syncretist of the *Oratio*," Grafton forcefully determines, "became a discriminating historical critic in the *Disputationes*." Anthony Grafton, "Giovanni Pico della Mirandola: Trials and Triumphs of an Omnivore," in *Commerce with the Classics: Ancient Books and Their Renaissance Readers* (Ann Arbor, Mich., 1997), 93–134, citation at 97. Finally, Brian Copenhaver has noted that post-Kantian scholars mistakenly portrayed Pico's *Oratio* as an essay about freedom and the dignity of man. Rather, he argues, it signifies an esoteric account of the ladder of philosophy, mystically climbing from one rung of the angelic sphere to another with the goal of achieving ecstatic union with God. Brian Copenhaver, "The Secret of Pico's Oration," *Midwest Studies in Philosophy* 26, no. 1 (2002): 56–81.

12. Horace (Ep. 1.1.14); Pico della Mirandola, *Oratio*, 30.

13. It is not a straightforward translation of Narbonni's commentary. Indeed, Narbonni's commentary is completely omitted, but since the starting paragraphs, where the words *dixit Abubacher* (said Abu-Baker Ibn-Tufayl) appeared only in the Hebrew translation, it is safe to assume that it was a translation of Narbonni's commentary. The handwriting in the marginalia matches the text itself and does not look like Pico's handwriting. A translator or a later copier most probably scribed the text and the marginalia. Most of the notes present alternative translations under the letters *AL*, which seem to be short for the Latin word *aliter* (otherwise). All told, it seems that this extant manuscript is a copy, probably made after Pico's death, of Pico's own copy, which has not survived.

14. Giovanni Pico della Mirandola, *Disputationes adversus astrologiam divinatricem*, ed. Eugenio Garin (Florence, 1946), 1:80.

15. Ibid., 80. In the original Latin, *How Anyone Can Become a Philosopher on His Own* reads "Librum quo quisque pacto per se philosophus evadat." I found "evadat" to be an unusual usage; it does not seem to have the meaning Garin gives it, as "sul modo in cui il filosofo raggiunge in liberta." Eugenio Garin translates *quisque* (each, every, any), and although *evado* can mean "to escape, to reach liberty," when it has that meaning it must be accompanied by a word or phrase to indicate what or who is escaping. Garin notes of "Albubather, Abubater, Abu Bakr al-Hasan ibn al-Khasib," "Of Persian origin, from the end of the ninth century, astrologer, wrote in Arabic and

in Persian. His best-known work is the *De nativitatibus* (Albubather Magni Alchasili Alcharsi filius *Liber de nativitatibus*) (Venice, 1492, 1501), translated into Latin by a clergyman, Salio di Padova, in 1218." Garin's footnote at 642.

16. In 1479 Niccolò Leoniceno, a professor of medicine and one of the leading medical humanists of the time, who used text-critical philological tools, especially his expert knowledge of Greek, to reform the study of medicine, was one of his masters in Ferrara. Leoniceno sparked in Pico the passion to work on Arabic texts. See, for instance, his translation of Galen from Greek to Latin. Niccolò Leoniceno, *In libros Galeni e Greca in Latinam linguam a se translatos prefatio communis* (Venice, 1508).

17. Delmedigo gave Pico private lessons in Hebrew, introduced him to Averroeist Latin translations, and brought the Kabbalah to his attention, though Delmedigo was reluctant to accept its seemingly irrational premises. Delmedigo composed an essay that attacked Kabbalah, *Beḥinat ha-dat: Examen religionis* (Jerusalem, 1969). Delmedigo ended his engagement with Pico very early in the 1480s, so he does not seem to be the one who introduced Pico to the story of Ḥayy Ibn-Yaqẓān.

18. Pico's most famous and prominent Hebrew teacher, who also extensively translated Kabbalistic writings for him, was Flavius Mithridates, a Sicilian Jew originally named Shmuel Ben-Nissim Abu al-Faraj before his conversion to Christianity. Mithridates introduced Pico to the corpus of Kabbalah during the late 1480s, but he, too, does not seem to be the one who introduced Pico to the story of Ḥayy, nor its translator. Chaim Wirszubski has traced the works that Mithridates engaged with Pico, and *Ḥayy Ibn-Yaqẓān* was not part of the list. Chaim Wirszubski, *Pico della Mirandola's Encounter with Jewish Mysticism* (Cambridge, Mass.: Harvard University Press, 1989). Moreover, by the early 1490s Pico and Mithridates had had a falling out and were on bad terms.

19. See Jochanan Alemanno, *Sha'ar ha-ḥeshek* (Livorno, 1790).

20. See B. C. Novak, "Giovanni Pico della Mirandola and Jochanan Alemanno," *Journal of the Warburg and Courtauld Institutes* 45 (1982): 125–47.

21. In the second part of the work Alemanno divides the universe into five ascending worlds: habit, imagination, thought, senses, and, at the pinnacle, defined knowledge. He stresses that sensual perception comes before cognitive knowledge about nature and suggests that the material world should be studied first. Jochanan Alemanno, *Ḥayy ha-'olamim*, Comunita Israelitica, Mantua, Ebr. 21.

22. Alemanno writes, "You should remember that during conception, the matter from which the embryo was created, will make a big bubble like a bubble that will come up from dough that was thrown into boiling oil. And that bubble will gradually become bigger when the embryo grows and it [the bubble] was made to protect the embryo and it is called placenta." Alemanno's marginalia as it appears on a copy of Narbonni's *Yeḥiel Ben-'Uriel*, Bayerische Staatsbibliothek, Munich, cod. Hebr. 59, 36a.

23. Narbonni, *Yeḥiel Ben-'Uriel*, 146a–146b. I compared the commentary with two other manuscripts of *Yeḥiel Ben-'Uriel* from the same period and place: Biblioteca Apostolica, at the Vatican, Hebr. 209, and also Biblioteca Comunale, Mantua, Hebr. 12. In the eighth and last part of Narbonni's commentary on Ḥayy's story, he discusses

Hayy and Absal's practice of repeating the names of God. In this part of the manuscript Alemanno has inserted capitalized words for each of the names of God, broken down into its constituent letters: מיוד**הא**ואו**הא. *Yehiel Ben-'Uriel*, Bayerische Staatsbibliothek, Munich, cod. Hebr. 59, 156b.

24. Pico della Mirandola, *Oratio*, 7.

25. Pico della Mirandola, *Heptaplus*, trans. Douglas Carmichael (Indianapolis: Hackett Publishing Company, 1998), 117.

26. Ibid., 135, 148, 150.

27. The scholarship is divided on the question of when Alemanno and Pico met. Arthur Lesley, who has written extensively on Alemanno, reckons that Pico and Alemanno did not meet before 1486. He rejects the possibility that Alemanno translated the text for Pico, because there is not a single word in Latin in any of his Hebrew writings, including the autographs, and his transliterations of non-Hebrew words are from Italian, in dialectal forms. See Arthur Lesley, *The Song of Solomon's Ascents by Yohanan Alemanno: Love and Human Perfection According to a Jewish Colleague of Giovanni Pico della Mirandola* (Ph.D. diss., University of California, Berkeley, 1976). On the other hand, Idel is one of the only scholars to argue that Pico and Alemanno met before 1488, even, perhaps, while Pico was writing the *Oratio*. For Idel, Pico and Alemanno were engaged with Hebrew translation of another Andalusian Islamic work of al-Baṭalyawsī (1052–1127), titled *Kitāb al-Ḥadaiq*, that describes the whole universe through the metaphor of a cosmic circle and the stages of creation descending from God. The work gave a special place to man mediating between God and nature. Alemanno and Pico, Idel argued, had different readings of this story. Whereas Alemanno placed man at the center of the universe, armed with a practical attitude toward positive altruistic activity, Pico stressed the actualization of individual potentialities as the ultimate *desideratum* of human perfection: that man can change his position by his acts. See Moshe Idel, "The Ladder of Ascension: The Reverberations of a Medieval Motif in the Renaissance," in *Studies in Medieval Jewish History and Literature*, ed. Isadore Twersky (Cambridge, Mass.: Harvard University Press, 1984), 2:83–93; and Moshe Idel, "The Anthropology of Yohanan Alemanno: Sources and Influences," *Topoi* 7, no. 3(1988): 201–10.

28. Abramo later settled in St. Marco, where he served as the first Hebrew teacher to the famous Hebraist Sante Pagnini, one of the first scholars to make biblical translations from sources other than the Vulgate translation of the Hebrew Bible. In 1521 he published a text of Psalms 1 to 28 in Latin, Hebrew, Aramaic, and Greek. Franco Bacchelli, "Pico della Mirandola traduttore di Ibn Tufayl," *Giornale critico della filosofia Italiana* 13 (1993): 1–25. Bacchelli compared the handwriting of Delmedigo and Mithridates to the handwriting of the manuscript and excludes them as possible translators (p. 8).

29. Franco Bacchelli, *Giovanni Pico e Pier Leone da Spoleto: Tra filosofia dell'amore e tradizione cabalistica* (Florence, 2001), 101.

30. Rutkin argues that Pico used astrological figurative language; for example, reflecting on himself, "There is an innate force (*vis*) greater than our mind (*animus*) / which denies that anyone lives by his choice (*arbitrio suo*)." Poem translated by Darrell

Rutkin in "Astrology, Natural Philosophy, and the History of Science, c. 1250–1700: Studies Toward an Interpretation of Giovanni Pico della Mirandola's Disputationes Adversus Astrologiam Divinatricem" (Ph.D. diss., Indiana University, 2002), ch. 5.

31. Carte 1, sec. 15 ex., Codici Gianni 46, Archivio di Stato, Florence. The names of the planets in this horoscope are written in words and not symbols, which may indicate Pico's lack of knowledge of astrology. After he died, Pico's horoscope, perhaps because of his anti-astrological tract, interested some astrologers, who even predicted his death. Luca Guarico (1476–1558), bishop of Civita, wrote about famous people's horoscopes, Pico's among them. Luca Guarico, *Tractatus astrologicum* (Venice, 1552). In his *Speculum astrologicum,* Francesco Guintini (1523–90) named Lucio Bellanti (who wrote against Pico's *Disputationes*), Antonio Sirigatti, and Angelo Catastini as three astrologers who had predicted Pico's death. Francesco Guintini, *Speculum astrologicum* (Lyons, 1573). See Patrizia Castelli, "L'oroscope di Pico," in *Pico, Poliziano e l'Umanesimo di fine Quattrocento,* ed. Paolo Viti (Florence, 1994), 225–29. Benivieni was author of *Dell'amore celeste e divino,* on which Pico wrote in his *Commento.* There, Pico also cites the Jewish astrologer Ibn-'Ezra (1093–1167), according to whom "Venus was placed in the middle of the sky near Mars in order that she might tame his violence." For Pico della Mirandola's commentary on *Dell'amore celeste e divino,* see *Commentary on a Poem of Platonic Love,* trans. Douglas Carmichael (London, 1986), 40.

32. Pico della Mirandola, *Oratio,* 37, 38, 38.

33. *Syncretism in the West: Pico's 900 Theses (1486),* trans. S. A. Farmer (Tempe, Ariz.: 1998), conclusion 9–3, 495; conclusion 10–21, 513.

34. Pico's conclusion 9–6 justifies the pursuit of self-directed education as allowed by Kabbalah and natural magic because God "pours the supercelestial waters of wondrous powers (*mirabiles virtutes*) daily on contemplative men of goodwill." Ibid., 497.

35. Furthermore, he claimed that other good Catholics, such as Albertus Magnus, used "*naturalis magia*" "that learned many things in nature by experiments." Pico della Mirandola, *Apologia,* in *Opera Omnia,* ed. Gian Francesco Pico della Mirandola (Hildesheim, Ger.: G. Olms, 1969), 1:169.

36. Ibid., 170.

37. There was, for example, the well-established tradition of astrological writings that predicted profound transformations in world politics in 1484. Then some predicted the coming conjunction of the planets that was supposed to take place in 1524. Finally, Pico's Florentine environment was particularly saturated with astrological influence. In the letter he sent on May 30, 1488, Ficino invited Pico to live under Lorenzo de' Medici's patronage and asserted their celestial bond: both were born when Saturn touched Aquarius, and "by Saturn it came about that Lorenzo the magnificent—among the Saturnine people the most worthy—took care of me and recalled Pico to Florence... He already commanded you when you first came to Florence to live under the great conjunction [of 1484]. So be then happy and Florentine." Marcilio Ficino, *Opera,* I:888; citation and translation from Darrell Rutkin, "Astrology, Natural Philosophy, and the History of Science," ch. 6, p. 3.

38. Pico della Mirandola, *Heptaplus*, 119.
39. Pico testifies in the *Disputationes* that he and his friends regularly made fun of the astrologers. Pico della Mirandola, *Disputationes*, 1:61.
40. Pico della Mirandola, "In suburbana mea villa, in qua haec scripsimus" in *Disputationes*, 1:162.
41. See Garin's introduction to Pico della Mirandola, *Disputationes*, 3-8. Italics mine.
42. Pico della Mirandola, *Disputationes*, 1:45-47.
43. "'Mitto etiam ad te,' scrive, 'graecum epigramma, quod effutivi, quod effutivi nuper, astrologis istis tuis iratus, qui cum professione istorum rixantem, diutius etiam quam velim, ruri te remorentur.'" Garin's introduction to Pico della Mirandola, *Disputationes*, 3-8.
44. As a result, it is impossible to know which parts of Pico's original drafts were screened or about some comments whose originality may be attributed to the numerous editors.
45. See Giorgio Radetti, "Un'aggiunta alla biblioteca di Pierleone Leoni da Spoleto," in *Rinascimento: Rivista della'Istituto Nazionale di Studi Sul Rinascimento* (Florence, 1965), 5:87-99.
46. "Angelo Poliziano to his dear Jacopo Antiquari," in *Angelo Poliziano: Letters*, letter 4:2, pp. 227-51. Quotations in the following three paragraphs are from the same source.
47. Ibid. Poliziano here made oblique reference to the rumor that one of Lorenzo's family members actually threw Pierleone down a well in an act of revenge. See Giorgio Radetti, "Un'aggiunta alla biblioteca di Pierleone Leoni da Spoleto," 88.
48. Ludovico Frati, "La morte di Lorenzo de' Medici e il suicidio di Pier Leoni," *Archivio storico Italiano* 5, no, 4 (1889): 255-60, 259.
49. Luigi Guerra-Coppioli, "Pierleone da Spoleto, medico e filosofo. Note biografiche con documenti inediti," in *Bollettino della R. Deputazione di storia patria per l'Umbria* (Perugia: Unione tip. coop., 1915), 21:430.
50. Machiavelli, *History of Florence*, bk. 8, ch. 7.
51. Pico della Mirandola, *Disputationes*, 1:447.
52. Christopher Fulton, "The Boy Stripped Bare by His Elders: Art and Adolescence in Renaissance Florence," *Art Journal* 56 (1997): 31-40, 31.
53. See Lorenzo Polizzotto, *Children of the Promise: The Confraternity of Purification and the Socialization of Youth in Florence, 1427-1785* (Oxford: Oxford University Press, 2004).
54. See Michael Rocke, *Forbidden Friendships: Homosexuality and Male Culture in Renaissance Florence* (Oxford: Oxford University Press, 1996), 155.
55. Serge Bramly, *Leonardo: Discovering the Life of Leonardo da Vinci* (New York: Edward Burlingame Books, 1991), 129.
56. See Rocke, *Forbidden Friendships*, 198n11.
57. Ibid., 201n19.
58. Ibid., 201.
59. During the two and a half years in which he ruled, Piero, Lorenzo's son, bru-

tally repressed sodomy. Whereas only eight people were convicted of sodomy each year between 1480 and 1491, during the period of April 1492 to February 1494 forty-four men were condemned and convicted. Ibid., 202–3.

60. Pico della Mirandola, *Disputationes*, 1:412. This story is cited from Grafton, 121.

61. In 1486 Pico corresponded with the humanist Ermolao Barbaro about "rhetoric," a coded reference to sodomy, complaining about long, tedious speeches—"long-haired speech is always sodomotical"—and using the Latin word *cinaedus*, which typically refers to a passive partner, usually a youth, in a homosexual relationship. *Renaissance Debates on Rhetoric*, ed. and trans. Wayne A. Rebhorn (Ithaca, N.Y.: Cornell University Press, 2000), 58n2.

62. Giovanni Pico della Mirandola, *Commento*, to G. Benivieni's *Canzone: Dell'-amore celeste e divino* (Luca, 1731), 83–84.

63. Quoted by Giovanni Dall'Orto, "'Socratic Love' as Disguise for Same-Sex Love in Italian Renaissance," in *The Pursuit of Sodomy: Male Homosexuality in Renaissance and Enlightenment Europe*, ed. Kent Gerard and Gert Hekma (New York, 1989), 43. From Isidoro del Lungo, *Florentia* (Florence, 1897), 277–78.

64. Leone Ebreo, *Dialoghi d'amore*, ed. Santino Caramella (Bari: G. Laterza and figli, 1929), 159.

65. Pico's attack on astrology in the *Disputationes* also contained a tone of frustration about false astrological prognostics regarding his siblings. The famous astrologer of the time, Girolamo Manfredi, promised Pino Ordelaffi, the signore of Flori and Pico's brother-in-law [married to his sister Lucretia] all sorts of wonderful things to happen in 1474; alas, the wonderful things not only did not occur, he died that year. Pico concludes, "Who will make good his guaranteed claims I know not how, unless perhaps he would look down on earthly things more truly from the heavens than he formerly looked up at the heavens from earth." Pico della Mirandola, *Disputationes*, 1:164. In 1491 the same thing happened with his sister-in-law, Constantia, who warned before her death of the astrologers' false predications. Pico writes in a bitter and frustrated tone about a recent event in which Constantia, his brother Antonio's wife, who was guaranteed favorable health by an astrologer, died the same year (1491), and with her last breath said to her husband, "Behold how true the prognostications of the astrologers are!" Pico della Mirandola, *Disputationes*, 1:166.

66. Girolamo Manfredi, *Libre de homine* (Bologna, 1474).

67. For a survey of astrological representations of homosexuality in early sixteenth-century astrological commentaries, mainly of Cardano, Schöner, and Pontano, see P. G. Maxwell-Stuart, "Representations of Same-Sex Love in Early Modern Astrology," in *The Sciences of Homosexuality in Early Modern Europe*, ed. Kenneth Borris and George Rousseau, 165–83 (London: Routledge, 2008); and Darrell Rutkin, "Astrological Conditioning of Same-Sexual Relations," ibid., 183–201.

68. Pico della Mirandola, *Disputationes*, 1:459.

69. B. C. Novak, "Giovanni Pico della Mirandola and Jochanan Alemanno," 140–41.

70. Pico della Mirandola, *Oratio*, 18.

71. Pico della Mirandola, *Commento*, 37.

72. Ibid., 4.

73. Vittore Branca, "Alla ricerca di libri e di docenti nel Vento umanistico," in *Poliziano: E l'umanesimo della parola* (Torino, 1983), 134–55.

74. They continued on to Padua, where they met Pierleone da Spoleto, who showed them his private library, where Poliziano was fascinated to find and purchased a copy of Marcus Manilius's *Astronomicon*, a Roman astrological poem in five books, prepared for publication by the astronomer Johannes Regiomontanus (1436–76) and first published in 1473. For most of the trip, Pico felt unwell and complained of pain in his eyes, but he recovered when they arrived in Venice, and they took a boat trip, visited the Bessarione library in the Basilica di San Marco and left for Florence with more than thirty additional volumes. Vittore Branca, "Alla ricerca di libri e di docenti nel Vento umanistico," 137.

75. "Niccolò Leoniceno to his dear Angelo Poliziano," in *Angelo Poliziano: Letters*, letter 2:7, p. 101.

76. "Angelo Poliziano to his dear Giovanni Pico della Mirandola," in *Angelo Poliziano: Letters*, letter 1:7, p. 26–27.

77. "Tum illud quod fecit nobis facile discoperire hoc secretum et abrumpere hoc velum ab eo quoniam videmus in hoc nostro tempore oppiniones coruptas quas dixerunt illi qui existimant se philosophos et declaraverint, usquequo extenderint se per terra cum magna fortitudine, et propter timorem nostrum eorum qui sunt debiles, qui proiecerunt cabala profetarum et voluerunt recipere cabla huius oppinionis et descriptionem eius, quoniam cogitaverunt quod ille oppiniones erant secreta que ymaginantur in eis ultra eos et auxerunt in hoc studium eorum et amorem suum in ea. Et vidimus nos splendere in aliqua parte de secretis secretorum per abstractionem ad veritatem et removere ab ea via . . . et ego supplico fratribus meis qui consistent in hac sermonicatione quod recipient narrationem veram in qua ego feci facile." Pico della Mirandola, *Ḥayy Ibn-Yaqẓān*, fols. 115v–116r.

78. Pico della Mirandola, *Commento*, 34.

79. Pico della Mirandola, *On Being and One: To Angelo Poliziano*, trans. Paul Miller (Indianapolis: Hackett Publishing, 1998), 37.

80. Marsilio Ficino, *Platonic Theology*, bk. 9, ch. 3, p. 2.

81. Ibid., bk. 6, ch. 2, p. 9.

82. Marcilio Ficino, *De vita libri tres* (a reprint of the 1489 edition published in Venice was published in New York by G. Olms Verlag, 1978), bk. 3, pp. 3–4.

83. Pico della Mirandola, *Disputationes*, 1:203.

84. Ibid., 1: 269.

85. "Dixit Abubachar. Nominaverunt nostri antiqui iusti quod inter insulas indie est una insula, que stat sub linea equatori set hec est insula in qua natus homo est sine patre et matre, quoniam ipsa est temperatioris aeris quam loca totius terre et perfectior eis propter illustrationem lucis super eis cum dispositione." Pico della Mirandola, *Hayy Ibn Yaqzan*, fol. 84v.

86. "Dixit Abubachar. Tamen illi quid cogitaverunt quod ipse natus sit sine patre et matre dixerunt quod venter terrae illius insulae fermentarat intra eam lutum in longitudine annorum, quousque complexionatus in eo humidum cum sicco et calidum cum frigido et fuit complexionatus cum actione passione et equalitate im potentia et

erat hoc magne fermentationis multum et erat excessus alterius super alteram cum equalitate complexionis dispositionis ad generationem medius inter humores magis temperatos quam sint in eis et perfectioris, similis complexioni humane et mixtum est ipsum lutum minimum unius cum minimo alterius et producte sunt in eo vessice ad similitudinem ampullarum ex ebullitione vi mixtionis facte et producta est in eius medio vessica una parva valde et dividitur in duas partes inter se distantes, altera plena corpore subtili aereo in fine ultimo temperamenti quod est ei conveniens et sequitur ad hoc spiritus qui est adeo benedictum—benedicitur ipse—fixavit eum fixione qua grave est separare ab eo esse sensum et esse intellectum. Ibid., fols. 85v–86r.

87. Pico della Mirandola, *Disputationes*, 1:162.

88. Ibid., 1:293.

89. Grafton, "Giovanni Pico della Mirandola: Trials and Triumphs of an Omnivore," 93–134.

90. In 1502 Jacopo Constanzo wrote about "a magic poisoned beverage" that Poliziano drank. See the collection of evidence on the death of Poliziano in Carlo Pionisotti, "Considerazioni sulla Morte del Poliziano," in *Culture et société en Italie: Du moyen-âge à la Renaissance* (Paris: Université de la Sorbonne Noubelle, 1985), 146–48.

91. Ibid., 150–53.

92. Pico, *Disputationes*, 1:417.

93. Gianfrancesco Pico della Mirandola, "The Life of Pico," trans. Thomas More, in *The Yale Edition of the Complete Works of Thomas More* (New Haven: Yale University Press, 1997), 1:71.

94. Quoted ibid., 1:74.

95. Girolamo Savonarola, "Aggeus, Sermon XIII: Third Sunday of Advent, 12 December 1494," in *Selected Writings of Girolamo Savonarola: Religion and Politics, 1490–1498*, ed. Donald Beebe, Alison Brown, and Giuseppe Mazzotta (New Haven, Conn.: Yale University Press, 2006), 151–52.

96. Ibid., 158.

97. Luca Landucci, "A Florentine Diary," in *Selected Writings of Girolamo Savonarola, : Religion and Politics, 1490–1498*, 210.

98. Girolamo Savonarola, "Letter from Paolo de Somenzi, Orator, to Lodovico Sforza, Duke of Milan," in *Selected Writings of Girolamo Savonarola: Religion and Politics, 1490–1498*, 211.

99. Girolamo Savonarola, "La vita del beato Ieronimo Savonarola," in *Selected Writings of Girolamo Savonarola: Religion and Politics, 1490–1498*, 212.

100. Franco Bacchelli makes a forceful argument that Fregoso had access to Pico's manuscript-translation of Ḥayy through a common friend. Fregoso was a good friend of Giovanni Simonetta's son, Bartolomeo, and Pico communicated with Bonifacio Simonetta, Giovanni Simonetta's cousin. Fregoso also kept in touch with Giovanni Bolognini Castellano di Pavia, whose son, Pietro, married Elenora, Pico's cousin. See Bacchelli, *Giovanni Pico e Pier Leone da Spoleto*, 101.

101. See Antonio Fileremo Fregoso, "De lo istinto naturale," in *Opere*, ed. Giorgio Dilemmi (Bologna, 1976), 308–23.

102. See Mario Biagioli, "Knowledge, Freedom, and Brotherly Love: Homosociality and the Accademia dei Lincei," *Configurations* 3, no. 2 (1995): 139–66. See also the printed edition Lincei statues and bylaws, *Lynceographum: Quo norma studiosae vitae lynceorum philosophorum exponitur,* ed. Anna Nicolò (Rome: Accademia Nazionale dei Lincei, 2001), 190.

103. Francis Bacon, "The Great Instauration," in *Selected Philosophical Works,* ed. Rose-Mary Sargent (Cambridge, Mass.: Hackett Publishing, 1999), 89.

104. Ibid., 89.

Chapter 4 · Employing the Self and Experimenting with Nature: Oxford, 1671

1. Thomas More, "De tristitia Chrisiti," in *Complete Works of St. Thomas More,* ed. Clarence H. Miller (New Haven, Conn.: Yale University Press, 1963–ca. 1997), vol. 14, part 1, 445 47. It is not surprising that in his attack on autodidacts, More lingers upon St. Jerome, who canonized the Latin translation of the scriptures and had to fight other scholars who did not accept his authorized translation. More frequently, he reproved those who thought that no study or learning are required to interpret scripture. See *Patrologiae cursus completes, series Latina,* ed. J. P. Migne, 22:544.

2. *Philosophical Transactions of the Royal Society* 6, no. 73 (1671): 2214.

3. Edward Pococke, *Philosophus autodidactus, sive, Epistola Abi Jaafar ebn Tophail de Hai ebn Yokdhan microform: In quâ ostenditur quomodo ex inferiorum contemplatione ad superiorum notitiam ratio humana ascendere possit/ex Arabicâ in linguam Latinam versa ab Edvardo Pocockio* (Oxford, 1671), A2.

4. Ibid.

5. Ibid.

6. In our own day, the pertinent historiography to the case of the publication of *Philosophus autodidactus* goes in two directions. The historiography of science has recognized the importance of local English foundations in the construction of the practice and philosophy of experimentalism. On the other hand, the historiography of orientalism mostly engaged with Enlightenment Europe's fascination with the East and its texts and objects. Some modern scholarship has started exploring the Enlightenment European fascination with Orientalia in general and with the story of Ḥayy Ibn-Yaqẓān in particular. It points to the influence of Ibn-Tufayl's work on the modern age in Europe; the connection to Defoe's *Robinson Crusoe* is seen as particularly strong. Gerald J. Toomer's erudite work on the study of Arabic in seventeenth-century England, especially at Oxford, is based on extensive research into manuscript sources and provides authoritative biographical and bibliographical information rather than a sociohistorical interpretation of the rise of English experimentalism. Gerald J. Toomer, *Eastern Wisdom and Learning: The Study of Arabic in Seventeenth-Century England* (Oxford: Oxford University Press, 1996). Most particular concerns were given to the reception of *Philosophus autodidactus* in European intellectual culture. G. A. Russell speculated a causal connection between the publication of *Philosophus autodidactus* and the completion of Locke's first draft of the *Essay Concerning Human Understanding* in July 1671. For Russell, Ḥayy's story "provides for Locke, a clear and

detailed demonstration of how scientific principles could be empirically acquired. At the same time it also constitutes a tailor-made case for Locke's conviction [in *Human Understanding*] that sensory knowledge formed the basis of 'reflection,' leading to the discovery of moral truth." G. A. Russell, "The Impact of *Philosophus Autodidactus*: Pococke, John Locke, and the Society of Friends," in *The "Arabick" Interest of the Natural Philosophers in Seventeenth-Century England* (New York: Brill, 1994). More recently, Samar Attar uses literary criticism to point to the many authors in the early modern Western tradition who may have drawn, wittingly or not, upon Ibn-Tufayl's work, heralding the modern age in Europe. Samar Attar, *The Vital Roots of European Enlightenment: Ibn-Tufayl's Influence on Modern Thought* (Lanham, Md., 2007). Shelly Ekhtiar notes that the work was received well in England and poorly in France. She suggests that the tract was popularized in England by several English translations, the most popular of which was by Simon Ockley (1708), who picked up Locke's arguments and popularized them. Shelly Ekhtiar, "*Ḥayy ibn Yaqẓān*: The Eighteenth-Century Reception of an Oriental Self-Taught Philosopher," *Studies on Voltaire and the Eighteenth Century*, no. 302 (2002): 217–45. Accenting the Arabic origin of European culture, Arab literary scholars like Ḥasan Maḥmūd Abbās propose that Defoe's *Robinson Crusoe* directly derives from Ḥayy's story as presented by Pococke in *Philosophus autodidactus* and was fitted to Locke's essay as a philosophical framework. Ḥasan Maḥmūd Abbās, *Ḥayy ibn Yaqẓān wa-Rūbinsūn Kirūzū: Dirāsah muqāranah* (Beirut: Al-Muassasah al-Arabīyah lil-Dirāsāt wa-al-Nashr, 1983).

7. *Early voyages and travels in the Levant*, edited with an introduction and notes by J. Theodore Bent, vol. 1, *The diary of Master Thomas Dallam, 1599–1600*; and vol. 2, *Extracts from the diaries of Dr. John Covel, 1670–1679. With some account of the Levant Company of Turkey merchants* (New York, 1964).

8. Charles Robson, *Nevves from Aleppo: A letter written to T. V. B. of D. Vicar of Cockfield in Southsex· By Charles Robson Master of Artes, fellow of Qu: Col: in Oxford, and preacher to the Company of our English Merchants at Aleppo. Containing many remarkeable occurrences obserued by him in his iourney. thither* (London, 1628), 14–15.

9. Mortimer Epstein, *The Early History of the Levant Company* (London, 1908), 161.

10. One example of the way economic practices bridged cross-cultural difficulties was a work of William Barret, a commercial agent and consul during the early years of the English trade. From his accumulated knowledge of the Levant spice trade, he compiled detailed notes, published under the title *The money and measures of Babylon, Balsara, and the Indies, with the customs, &c. written from Aleppo in Syria, An. 1584*, in which he supplied a conversion system for terminology, measures, and prices. Such notes were essential information for any successful trader. Barret also remarked on the seasonal winds and monsoons and the associated appropriate sailing dates. Considering that cross-cultural accounts often strived to make the interactions between two cultures more efficient, this first cross-cultural index bridging the Near East and England was produced out of economic necessity. Richard Hakluyt, [1552–1616], *The principal navigations, voyages, traffiques & discoveries of the English nation made by sea or over-land to the remote and farthest distant quarters of the earth at any time within the compasse of these 1600 yeeres* (New York, 1969), 6:10–34. See

also the entry of William Barret in the *Oxford Dictionary of National Biography,* s.v. Barret, William. The Near East shaped English economic careers as well as political and cultural ones. Paul Pindar, a merchant and diplomat, became involved in commercial diplomacy as the consul of the English merchants in Aleppo and, after 1611, served as the English ambassador in Constantinople. Through his experience in commercial diplomacy, he accumulated wealth that was later used during the Civil War, supplying weapons in support of King Charles, especially during the fighting around Oxford. Pindar's Levantine wealth helped him pursue new cultural interests. Thomas Bodley, the founder of the Bodleian Library, called upon Pindar's connections in the East—and culled from Pindar's treasures—to establish his library. Bodley had a vision that the library would extend beyond European languages, with the ultimate goal of building up an impressive collection of Near Eastern materials. During the second decade of the seventeenth century he obtained works in or on Hebrew, Syriac, Arabic, Turkish, and Persian through such benefactors as Paul Pindar. Thomas Bodley, *The life of Sir Thomas Bodley, the honourable founder of the publique library in the University of Oxford, written by himselfe* (Oxford, 1647).

11. The first publication of 1740 was attached to Pococke's collection of Christian writings. See *The theological works of the learned Dr. Pocock, sometime professor of the Hebrew and Arabick tongues, in the University of Oxford, and canon of Christ-Church: containing his Porta Mosis, and English commentaries on Hosea, Joel, Micah, and Malachi. To which is prefixed, an account of his life and writings, never before printed; with the addition of a new general index to the commentaries. In two volumes,* ed. Leonard Twells (London, 1740). In the early nineteenth century the biography was published in Twells, *The Lives of Dr. Edward Pocock, the celebrated orientalist, by Dr. Twells; of Zachary Pearce, bishop of Rochester, and of Dr. Thomas Newton, bisiop of Bristol, by themselves; and of the Rev. Philip Skelton, by Mr. Burdy,* 2 vols. (London: R. and R. Gilbert for F. C. and J. Rivington, 1816), 1:1–356.

12. Laud instructed Pococke and others, such as John Greaves, to collect manuscripts and coins. See Anthony Wood, *Athenæ Oxonienses.* . . . (Oxford: Ecclesiastical History Society, 1848), vol. 2, col. 157. Thoma Smitho, *Vita quorundam eruditissimorum et illustrium virorom: quorum nomina exftant in pagina fequenti* (London: D. Mortier, 1707), 7.

13. See also descriptions of other contemporary Europeans in the Near East, such as Harlay de Cèsy, the French ambassador in Istanbul, who collected local works as part of French-Ottoman gift exchange. Gèrard Tongas, *Les relations de la France avec l'empire ottoman durant la première moitié du xviie siècle et l'ambassade à Constantinople Philippe de Harlay, comte de Cèsy, 1619–1640* (Toulouse: Impr. F. Boisseau, 1942). See also the description given by Italian traveler Pietro della Valle, *De viaggi di Pietro Della Valle Il pellegrino: Descritti da lui medesimo in lettere familiari all'erudito suo amico Mario Schipano* (G. Gancia, 1843), 1:153–55.

14. In a letter to Locke Pococke writes that he was awaiting Robert Boyle's "Commands and Instructions for the right improvement of my time in these Parts." See *Correspondence of John Locke,* ed. E. S. De Beer (Oxford: Clarendon Press, 1976–989), 1:353, letter 253.

15. Twells, *The Lives of Dr. Edward Pocock*, 30.
16. Twells found the letter Edward Pococke Jr. translated after his father's death. See Twells, *The Lives of Dr. Edward Pocock*, 31–33.
17. Ibid., 1:15; 1:16; 1:15.
18. Ibid., 1:17–18.
19. See for instance, Bochart's endeavor to solve these inconsistencies. Samuel Bochart, *Hierozoicon sive bipertitum opus de animalibus S. Scripturæ: cuius pars prior libris IV. De animalibus in genere & de quadrupedibus, viviparis & oviparis. Pars posterior libris VI. De avibus, serpentibus, insectis, aquaticis, & fabulosis animalibus agit* (London, 1663), bk. 1, ch. 9.
20. Pococke writes in 1677, "To prevent such mistake . . . a rule may be thus summed up, that wheresoever we meet with *Tannim* as plurals, they signify *those howling wild beasts inhabiting waste desolate places*. But where we meet *Tannim* in the singular they are to be rendered, *Dragons*, or *Serpents*, or *Sea-monsters*, or *Whales*, or the like." Edward Pococke, *A Commentary on the Prophecy of Micah by Edward Pocock* (London, 1677), ch. 1, verses 5–8.
21. See Twells, *The Lives of Dr. Edward Pocock*, 29.
22. Ibid., 41.
23. Letter cited ibid., 128.
24. Menashe Ben-Israel, Spinoza's teacher and a leader of the Jewish community in Amsterdam, renowned for his appeal to Cromwell to readmit the Jews to England, engaged in the search for them. In fact, an anonymous friend wrote to Pococke that Ben-Israel found among the Sephardic Jews of Amsterdam a learned Jew, linguistically skilled in Hebrew and Arabic, but his candidacy never advanced. Twells, *The Lives of Dr. Edward Pocock*, 138.
25. "As having travelled abroad, and been trained up, for many years, in the midst of those tongues and nations; that he hath been very useful here, and a great ornament to this University, where we understand he desires still, in all peaceable manner, to continue to serve this state, and his own country in this employment." Twells, *The Lives of Dr. Edward Pocock*, 137.
26. "When it is dried and thoroughly boiled, it allays the ebullition of the blood, and it is good against the smallpox and measles, and bloody pimples; yet it causes vertiginous headache . . . and assuages lust, and sometimes breeds melancholy . . . Some drink it with milk, but it is an error, and such as may bring in danger of the leprosy." Da'ud ibn 'Umar Antaki, *The nature of the drink kauhi, or coffe, and the berry of which it is made described by an Arabian phisitian* (Oxford, 1659), 3. In a 1659 letter to Boyle, Samuel Hartlib expressed excitement about this text and sought to promote Pococke's interests. He further strengthened Boyle's good impression of Pococke with the hope that it would expand the scale of interaction and business between them. "I thank you heartily for the printed paper on coffee, which will be gustful no doubt to your coffee drinkers." *Correspondence of Robert Boyle, 1636–1691*, ed. Michael Hunter, Antonio Clericuzio, and Lawrence M. Principe (Pickering, U.K., 2001), 1:327.
27. Pococke was one of a few scholars to whom Boyle often turned to translate oriental works and inscriptions. On Boyle's connection to Pococke as well as his activ-

ity in attempts to convert Muslims in the Near East, see Twells, *The Lives of Dr. Edward Pocock*, 1:242–45; 1:270; 1:276. Boyle himself wrote a preface for the first Turkish translation of the New Testament made by William Seaman, a colleague of Pococke's. Ibid., 1:276.

28. François Laplanche, *L'Evidence du Dieu Chrétien: Religion, culture et société dans l'apologétique protestente de la France classique* (1576–1670) (Strasbourg, 1983).

29. Hugo Grotius, *De veritate religionis christianae* (Paris, Sumptibus Seb. Cramoisy, 1640), pt. 2, section xviii, 50.

30. Grotius to W. de Groot, 16 February 1641, in *Briefwisseling van Hugo Grotius*, ed. P. C. Molhuysen, B. L. Meulenbroek, P. P. Witkam, H. J. M. Nellen, and C. M. Ridderikhoff, vol. 21, letter 5061 (The Hague; M. Nijhoff, 1928–2001), 103: "Fuit apud me his diebus Anglus vir doctissimus, qui diu in Turcico vixit imperio, et meum librum de veritate religionis christianae in Arabicum vertit sermonem; curabitque si potest typis in Anglia edi. Is nullum librum putat esse utiliorem aut instruendis illarum partium Christianis, aut etiam convertendis Mahumetistis, qui sunt in Turcico imperio, aut Persico, aut Tartarico, aut Punico, aut Indiano." See also J. P. Heering, *Hugo Grotius as Apologist for the Christian Religion: A Study of His Work "De veritate religionis christianae" (1640)* (Leiden, Neth.: Brill, 2004), 239.

31. Edward Pococke, *The Book of Common Prayer: Liturgiae Ecclesiae Anglicanae, partes praecipuae viz. preces matutinae & vespertinae; ordo administrandi coenam Domini; ordo baptismi publici; una cum ejusdem Ecclesiae doctrina, triginta novem Articulis comprehensa. Nec non homiliarnm [sic] argumentis: in linguam Arabicam traductae* (Oxford, 1674).

32. *Correspondence of Robert Boyle*, 2:53; 1:427; 1:450.

33. Ibid., 4:359–60.

34. Boyle's postscript informs Huntington of the publication of his new piece, probably *Saltness of the Sea*, evidence that cutting-edge books on natural philosophy circulated outside of Europe and reached the Near East. The Levant Company, then, played a role in circulating various printed texts—not only tools of conversion but also books on the new sciences. Ibid., 4:373.

35. Hugo Grotius, *De Veritate Religionis Christianae* (Paris, Sumptibus Seb. Cramoisy, 1640), pt. 2, section vi, 34; see also pt. 3, section vii, 65.

36. *Correspondence of Robert Boyle*, 3:95.

37. Ibid.

38. John Worthington, *The Diary and Correspondence of John Worthington* (Manchester, U.K.: 1847) 1:76.

39. The first mention of Hugo Grotius in Boyle's correspondence comes from a surprising direction. In September of 1659, John Evelyn wrote to Boyle, trying to convince him to get married. "I cannot consent, that such a person as Mr. Boyle be so indifferent, decline a virtuous love, or imagine, that the best ideas are represented only in romance, where love begins, proceeds, and expires in the pretty tale, but leaves us no worthy impressions of its effects. We have nobler examples: and the wives of philosophers, pious and studious persons, shall furnish our instances: . . . the late adventure of madam Grotius, celebrated by her Hugo who has not heard of?" Evelyn's

reference to the story about Grotius's wife in helping him to escape prison addressed Boyle's celibacy, a point discussed in his social network, which only gingerly touch upon his suspected homosexuality. *Correspondence of Robert Boyle*, 1:373.

40. For the printed edition of this manuscript, as well as for its context, see *Robert Boyle By Himself and Friends: With a Fragment of William Wotton's "Lost Life of Boyle,"* ed. Michael Hunter (London, 1994), xv–xxi. For the text itself, see *An Account of Philaretus During his Minority*, ibid., 1–22.

41. Boyle, *An Account of Philaretus During his Minority*, 3. Even the names chosen bear some similarity. The name Philaretus had a common meaning as "the person keen on being good." Boyle could have taken that name from a variety of works. Stories of Turkish history and exploits, in which the story of Byzantine Phileratus is related, abounded. Richard Knolles, *The generall historie of the Turkes from the first beginning of that nation to the rising of the Othoman familie: With all the notable expeditions of the Christian princes against them; Together with the liues and conquests of the Othoman kings and emperours faithfullie collected out of the-best histories, both auntient and moderne, and digested into one continuat historie vntill this present yeare 1603* (London, 1603), 11; Andrew Moore, *A compendious history of the Turks: Containing an exact account of the originall of that people; The rise of the Othoman family; and the valiant undertakings of the Christians against them: with their various events* (London, 1659), 6. In addition, the name Philaretus served as a "pseudonym" for writers on science, medicine, cultural entertainment, and theater. It appears as the pseudonym of a writer of a Latin medical reference book, translated from the Greek and titled *De pulsibus* (1567). Philaretus, *De Pulsibus*, in *Aetii medici graeci*, ed. Ianum Cornarium (London: Medical and Chirugical Society, 1567). In turn, Simeon Partlicius mentioned this chapter in *A new method of physick, or A short view of Paracelsus and Galen's practice* (London, 1654), 27. The name was also associated with the culture of personal habits and entertainment. In 1602 a booklet titled *Work for Chimny-Sweepers, or A Warning for Tabacconists* was published in London. The author was a doctor who styled himself Philaretus, and his writing was the first publication to present the risks of tobacco, calling for users to give it up and return to a healthy way of life. Philaretes, *Work for Chimny-Sweepers or A Warning for Tabacconists* (London: Thomas East for Thomas Bushell, 1602), cited in Anne Charlton, "Tobacco or Health 1602: An Elizabethan Doctor Speaks," *Health Education Research* 20, no. 1 (2005): 101–11. A character named Young Philaretus also appeared in the play *The Careless Shepherdess*, written by Thomas Goffe. Thomas Goffe, *The careles shepherdess a tragicomedy acted before the King & Queen, and at Salisbury-Court, with great applause, written by T.G.; with an alphebeticall catologue of all such plays that ever were printed* (London, 1656), 2. Boyle was also intellectually indebted to George Starkey's work in chemistry, and Starkey often published as "Eirenaeus Philalethes" (Lover of the Truth). William Newman and Lawrence Principe, *Alchemy Tried in the Fire: Starkey, Boyle, and the Fate of Helmontian Chemistry* (Chicago: University of Chicago Press, 2002).

42. Pococke apparently knew about Boyle's autobiography and the emblematic name Philaretus. In 1683 Grotius's polemic *De veritate* was translated into English by

Pococke's friend, Richard Baxter, who in fact veiled himself behind the name Philaretus. Doing so allowed Baxter to strategically gain Boyle's favor and patronage, indicating that in the circle of scholars surrounding Boyle, his pseudonym was used to curry favor. Philaretus, *Anti-Dodwellisme being two very curious tracts formerly written by the renowned Hugo Grotius, containing a solution of these two questions, I. whether the Eucharistie may be administred in the absence or want of pastours? II. whether it be necessary at all times to communicate with the symbols?; made English by Philaretus* (London, 1683). Baxter, who was most probably familiar with Boyle as Philaretus, wrote to him in June 1665 expressing his admiration for Boyle's publications, among them "the noble design of your Arabic publication of Grotius." *Correspondence of Robert Boyle,* 2:473.

43. Robert Boyle, "A Fragment of *The Aspireing Naturalist* (A Philosophical Romance)," *Boyle Papers,* vol. 8, fols. 206–7, Boyle Collection, Royal Society Archive, National Archive, Surrey, U.K.

44. Averroes has transmitted Avempace's view, according to which motion is measured by subtracting "from the motion only the retardation affecting it by reason of the medium, and its natural motion would remain." *Opera Aristotelis . . . cum Averrois commentaries* (Venice,1560), vol. 4, fol. 131v.

45. See Ernst Moody, "Galileo and Avempace: The Dynamics of the Leaning Tower Experiment," pt. 1, *Journal of the History of Ideas* 12, no. 2 (1951): 163–93; and "Galileo and Avempace: The Dynamics of the Leaning Tower Experiment," pt. 2, *Journal of the History of Ideas* 12, no. 3 (1951); Ernest A. Moody, "Empiricism and Metaphysics in Medieval Philosophy," *Philosophical Review* 67, no. 2 (1958): 161.

46. Pocoke, *Philosophus autodidactus,* A2.

47. Boyle's work takes the form of a dialogue with Pyro. Arguing against Epicurus and Lucretius's views about the randomness of events, he writes, "I oppose their arguments who justly deny the emergency of the world and especially of those animated bodies that help to compose it, from a casuall concourse of matter, which being on all hands conffes'd to consist of Atoms or particles inanimate whilst such, I confesse I can by noe means conceive Engines as Animalls plants etc. without guidance and conduct of an intelligent cause." Robert Boyle, "Essay on Spontaneous Generation," in *The Works of Robert Boyle,* ed. Michael Hunter and Edward B. Davis (London, 1999–2000), 13275, 280.

48. "When the wise Author of Nature had fashioned the universal matter into the world . . . and made the protoplasts or first individuals of each kind of living creatures . . . he fit in certain portions of matter he did foresee and design." Ibid., 287.

49. Peter Anstey, "Boyle on Seminal Principles," *Studies in History and Philosophy of Biological and Biomedical Siences* 33C, no. 4 (2002): 597–630.

50. The sixteenth-century physician Levinus Lemnius (1505–68) published *De miraculis occultis naturae* in Antwerp, in 1559, which dealt with the philosophical rules of "how man shall become excellent in all conditions, whether high or low, and lead his life with health of body and mind." Translated into English as *The Secret Miracles of Nature* (London, 1658). Daniel Sennert, "doctor of physick," and Nicholas Culpeper, "physitian and astrologer," wrote an extensive *Thirteen books of natural phi-*

losophy sometime around 1640, but it was not published until 1660. Book 5 is called "Concerning the spontaneous generation of live things." *Thirteen books of natural philosophy . . . By Daniel Sennert, doctor of physick. Nicholas Culpeper, physitian and astrologer. Abdiah Cole, doctor of physick, and the liberal arts* (London, 1660).

51. Pocock, *Philosophus autodidactus*, A2.

52. Robert Boyle, *The Christian virtuoso: Shewing that by Being Addicted to Experimental Philosophy, a Man is Rather Assisted than Indisposed to be a Good Christian, by T.H.R.B., Fellow of the Royal Society; to which are subjoyn'd, I. A Discourse about the Distinction that represents some things as above reason, but not contrary to reason, II. the first chapters of a discourse entitled, "Greatness of mind promoted by Christianity, by the same author"* (London: Edw. Jones for John Taylor, 1690).

53. The second part further adds to the confusion surrounding this diffuse and difficult book, for it remained among Boyle's unpublished papers at the time of his death and was not issued until it appeared in the fifth volume of Birch's first collected edition of Boyle's writings in 1744. As published in the collection of Boyle's works, it includes a long appendix to part 1, "concerning some papers belonging to the second part of the Christian Virtuoso . . . probably sent to his friend H.O. [Henry Oldenburg] Esq." Boyle, *The Works of Robert Boyle*, 12:736. As Boyle elsewhere testified in the preface, it was written many years beforehand, "thrown aside among other papers for several years." Ibid., 11:283.

54. Mario Sina, "Testi teologico-filosofici Lockiani dal ms. Locke c. 27," *Rivista di filosofia neoscolastica* 64 (1972): 54–75; M. A. Stewart, "Locke's "Observations" on Boyle," *Locke Newsletter* 24 (1993): 21–34.

55. Boyle, "The Christian Virtuoso," *The Works of Robert Boyle*, 11:292.

56. "Most of the perplexing Difficulties the Atheists lay so much stress on, proceed from the Nature of things; that is, partly from the Dimness and other Imperfections of our Human understandings, and partly from the Abstruse Nature, that, to such Bounded Intellects, all Objects must appear to have, in whose Conception infinity is involv'd; whether that Object be God, or Atoms, or Duration, or some other things that are uncausable. For, however we may flatter our selves, I fear we shall find, upon strict and impartial Tryal, that finite Understandings are not able clearly to resolve such Difficulties, as exact a clear comprehension of what is really Infinite." Boyle, "The Christian Virtuoso," *The Works of Robert Boyle*, 11:294.

57. "And this is strongly confirm'd by Experience, which witnesseth, that in almost all Ages and Countries, the generality of Philosophers, and contemplative Men, were persuaded of the Existence of a Deity, by the consideration of the *Phaenomena* of the Universe; whose Fabrick and Conduct they rationally concluded could not be deservedly ascrib'd, either to blind Chance, or to any other Cause than a *Divine Being*." Boyle, "The Christian Virtuoso," *The Works of Robert Boyle*, 11:295.

58. Keith was associated for some time with the Cambridge Platonist circle and especially with Henry More. His attraction to *Ḥayy Ibn-Yaqẓān* seems to be a follow-up to his key work, *Immediate Revelation* (1668), in which he claimed to prove that direct revelation was still possible by the inner light. This represents one of the Quakers' earliest systematic writings, in line with their emphasis on emotive experience

and practices rather than doctrinal systematic experience. See George Keith, *Immediate revelation, or Jesus Christ the eternall Son of God revealed in man microform: and revealing the knowledge of God and the things of his kingdom immediately: or, the Holy Ghost, the Holy Spirit of promise, the spirit of prophecy poured forth and inspiring man and inbuing him with power from on high . . . not ceased, but remaining a standing and perpetual ordinance in the Church of Christ and being of indispensible necessity as to the whole body in general. . . .* (Aberdeen, 1668).

59. *An account of the Oriental Philosophy Shewing the Wisdom of some Renowned Men of the East; and particularly the profound Wisdom of Hai Ebn Yokdan, both in Natural and Divine Things; Which he attained without all Converse with Men, (while he lived in an Island a solitary life, remote from all Men from his infancy, till he arrived at such perfection) Writ originally in Arabick, by Abi Jaaphar Ebn Tuphail; and out of the Arabick Translated into Latine by Edward Pococke, Student in Oxford and now faithfully out of his Latine, Translated into English: For a general service 1674*, trans. George Keith (London, 1674), 1.

60. Anna Ruth Fry, *Quaker Ways* (London: Cassell and Co., 1933), 203; Arthur Raistrick, *Quakers in Science and Industry* (London: Bannisdale Press 1950), 221–22; Frederick Tolles, *Quakers and the Atlantic Culture* (New York: Macmillan, 1960), 62–67; T. L. Underwood, "Quakers and the Royal Society of London in the Seventeenth Century," *Notes and Records of the Royal Society of London* 31, no. 1 (1976): 133–50.

61. See Henry Richard Fox Bourne, *The Life of John Locke* (London, 1876) 1:56, 88.

62. We get an indication of Locke's intensified interest in the Near East from a 1617 letter sent by Pococke's student, Robert Huntington, from Aleppo. Huntington describes his techniques of navigation and longitudinal determination and the disastrous effects of earthquakes: "The country is miserably decay'd, and hath lost the Reputation of its Name, and mighty stock of Credit it once had for Eastern Wisedome and learning: it hath followed the Motion of the Sun, and is Universally gone Westward." *Correspondence of John Locke*, vol. 1, p. 353, letter 253.

63. Mishnah, "Pirkei Avot," 4:20.

64. His son Edward (baptized 1648, d. 1726) took the same course of interest in oriental languages combined with clergy work. He was admitted as a student to Christ Church, where John Locke (who became a long-standing friend) acted as his tutor. After graduating in 1668, Pococke Jr. journeyed to Leiden, where in 1669 he was greeted by his father's admirers. After the publication of *Philosophus autodidactus* he followed his ancestors' footsteps and took ordination in 1672, embarking on an ecclesiastical career. According to Twells, Locke wrote that "he had spoken with Mr. Boyle about it [the translation of Maimonides]; he desired also to have it printed." Twells, *The Lives of Dr. Edward Pocock*, 291.

65. John Locke, *Essay Concerning Human Understanding* (London, 1689), bk. 1, ch. 3, art. 23; bk. 1, ch. 1, art. 27; bk. 1, ch. 2, art. 20; bk. 1, ch. 1, art. 27.

66. Ibid., bk. 1, ch. 2, art. 22; bk. 1, ch. 3, art. 2.

67. Ibid., bk. 1, ch. 3, art. 11.

68. The younger Pococke once wrote an apology to Locke for not thanking him earlier for the dedicated copy he had sent of a book, most probably a copy of Locke's

Essay Concerning Human Understanding, which had come out in print a year earlier: "I am commanded by my father to beg your pardon for his not writing at all. I doubt not but his age and infirmitys will be a good plea to you, who are so well acquainted with them; yet he is heartily sorry that he cannot let you know what value he has for your present, but by another hand. For my self Sir . . . I own it as the greatest obligation, that you are pleased at any time to have me in your thoughts; but shall not be able to understand how far I am indebted to you for your book, till I have persued it." *Correspondence of John Locke,* vol. 4, pp. 37–38, letter 1267. The full text of the eulogy appears in Twells, *The Lives of Dr. Edward Pocock,* 349–53, citation at 350.

69. Twells, *The Lives of Dr. Edward Pocock,* 350.
70. Ibid., 351.
71. John Locke, *Some Thoughts Concerning Education* (London, 1693): A4–A5.

Conclusion · Sampling the History of Autodidacticism

1. See, for instance, the conclusive work of Emanuel Frank, *Utopian Thought in the Western World* (Cambridge, Mass.: Belknap Press of Harvard University Press, 1979).
2. Isaiah Berlin, *Crocked Timber of Humanity* (New York: Knopf, 1990), 23, 29.
3. Giambattista della Porta, *Magia naturalis* (Naples, 1558), 21.
4. *Lynceographum: Quo norma studiosae vitae lynceorum philosophorum exponitur,* ed. Anna Nicolò (Rome: Accademia Nazionale dei Lincei, 2001), 5.
5. Horace "Ode on Endurance," bk. 3, ode 2.
6. For instance, the scrolls of the people in Bensalem, the city of New Atlantis, were written in Hebrew and Greek, signed by the Cherubim, two little angels with baby faces, positioned above a Jewish altar that had great import for Kabbalist commentators, who conceived it as connecting the high priest to the angelic world through his practice of sacrifices. There are also many indications that the *New Atlantis* is located in a continuum with the ancient Near East. The people in the *New Atlantis* had a hat resembling the Turkish turban; they called the island Bensalem, combining the Hebrew *ben* with the Arabic *salem* to render a phrase meaning "son of peace"; the house of science is called the house of Solomon, echoing the biblical figure who was able to speak the language of animals; in addition to the natives of the island there were also Hebrews, Persians, and Indians and frequent visits by Persians, Chaldeans, and Arabians. And then there is Bacon's most striking point, that "Moses by the secret cabala ordained the laws of Bensalem which they now use," tracing the origins of the society all the way back to the Bible itself.

ESSAY ON SOURCES

Versions and Editions of Ḥayy Ibn-Yaqẓān

From the twelfth century to the present, the story of *Ḥayy Ibn-Yaqẓān* has been vigorously circulated across cultures, enthusiastically translated into various languages, and intriguingly reincarnated into other literary forms. The major samples, listed below, indicate the strong reception and sustained appeal the story has had on generations of writers who draw on Ḥayy as the quintessential autodidact.

Abbās, Ḥasan Maḥmūd. *Ḥayy ibn Yaqẓān wa-Rūbinsūn Kirūzū: Dirāsah muqāranah.* Beirut, 1983.

Alemanno, Yochanan. *Yehiel Ben-Uriel.* Bayerische Staatsbibliothek, Munich. Cod. Hebr. 59.

Ashwell, George, trans. *The History of Hai Eb'n Yockdan, an Indian prince, or The self-taught philosopher . . . written originally in the Arabick tongue by Abi Jaafar Eb'n Tophail . . . ; set forth not long ago in the original Arabick, with the Latin version by Edw. Pocock. . . .* London, 1686.

Boigues, Francisco Pons, trans. *El filosofo autodidacto de Abentofail: Novela psicologica con un prologo de Menendez y Pelayo.* Saragossa, 1900.

Brönnle, Paul. *"The Awakening of the Soul": Rendered from the Arabic, with introduction, by Paul Brönnle.* London, 1905.

Gauthier, Léon, trans. *"Ḥayy ben Yaqdhan": Roman philosophique; Texte arabe, publié d'après un nouveau manuscrit . . . et traduction française.* Algiers, 1900.

Goodman, Evan Lenn, trans. *Ibn Tufayl's "Ḥayy ibn Yaqẓān": A Philosophical Tale.* New York, 1972.

Ibn Ṭufayl, Muḥammad ibn 'Abd al-Malik. *Het leeven van "Hai ebn Yokdhan."* Amsterdam, 1672.

———. *Risālat Ḥayy ibn Yaqẓān.* Bodleian Library, Oxford University. Pococke 263.

———. *Risālat "Ḥayy ibn Yaqẓān" fī asrār al-ḥikmah al-mashriqīyah: Istakhlaṣahā min durar jawāhir alfāẓ al-ra'īs Abī 'Alī ibn Sīnā . . . Abū Ja'far Ṭufayl.* Cairo: Maṭba'at Idārat al-Waṭan, 1882.

———. *On a Desert Island: "Ḥayy ibn Yaqzan," by Ibn Tufayl.* Abridged and adapted by Salah Abdul Sabour. Illustrated by Mostafa Hussein. Translated by Denys Johnson-Davies. Cairo: Dar el-Shorouk, 2002.

———. *Le philosophe sans maître: Histoire de "Hayy ibn Yaqzân."* Introduction by Georges Labica. Translated by Léon Gauthier. Algiers: S.N.E.D., 1969.

Keith, George, trans. *An account of the Oriental Philosophy Shewing the Wisdom of some Renowned Men of the East; and particularly the profound Wisdom of Hai Ebn Yokdan, both in Natural and Divine Things;... Writ originally in Arabick, by Abi Jaaphar Ebn Tuphail; and out of the Arabick Translated into Latine by Edward Pocok, Student in Oxford and now faithfully out of his Latine, Translated into English: For a general service 1674.* London, 1674.

Lafontaine, August Heinrich Julius. *Der Naturmensch, Oder, Natur und Liebe.* Vienna, 1799.

Mirandola, Giovanni, Pico della. *Ḥayy Ibn-Yaqẓān.* Biblioteca Universitaria di Genova. Cod. A, IX, ms. 29, 79v–116r.

Narbonni, Moshe. *Yehiel Ben-'Uriel.* Bayerische Staatsbibliothek, Munich. Cod. Hebr. 59.

Ockley, Simon, trans. *The improvement of human reason, exhibited in the life of Hai ebn Yokdhan: Written in Arabick above 500 years ago, by Abu Jaafar ebn Tophail... newly translated from the original Arabick, by Simon Ockley.* London, 1708.

Pococke, Edward. *Philosophus autodidactus, sive, Epistola Abi Jaafar ebn Tophail de Hai ebn Yokdhan in quâ ostenditur quomodo ex inferiorum contemplatione ad superiorum notitiam ratio humana ascendere possit, ex Arabicâ in linguam Latinam versa ab Edvardo Pocockio.* Oxford, 1671.

Pritsius, Georg. *Der von sich selbst gelehrte Weltweise, oder eine angenehme und sinnreiche Erzählung der wunderbaren Begebenheiten Hai Eben Jockdahns.* Frankfurt, 1726.

Schaerer, Patric O., trans. *Der Philosoph als Autodidakt: Ḥayy ibn Yaqẓān; Ein Philosophischer Inselroman.* Hamburg, 2004.

Major Writings of Custodians and Readers of Ḥayy Ibn-Yaqẓān

Key intertexualities and approximate connections can be found between the story of *Ḥayy Ibn-Yaqẓān* and the extensive intellectual production of its custodians and readers. The text circulated around the time of the production of major masterpieces that fostered liberalism, experimentalism, and republicanism. It has been echoed by various writers on a variety of subjects—from Pico della Mirandola's *On the Dignity of Man* through John Locke's *Essay Concerning Human Understanding* up to Jean-Jacques Rousseau's *Émile*—placing autodidacticism at the heart of modern thought.

Alemanno, Yochanan. *Sha'ar ha-ḥeshek.* Livorno, 1790.

———. *Hayy ha-'Olamim.* Comunita Israelitica, Mantua. Ebr. 21.

Boyle, Robert. *Correspondence of Robert Boyle, 1636–1691.* Edited by Michael Hunter, Antonio Clericuzio, and Lawrence M. Principe. 6 vols. London: Pickering and Chatto, 2001.

Bourne, H. R. Fox. *The Life of John Locke*. London, 1876.
Delmedigo, Joseph. *Sefer melo hofnayim*. Edited by Avraham Gaiger. Berlin, 1860.
———. *Sefer Elim*. Amsterdam, 1629.
Fregoso, Antonio Fileremo. "De lo istinto naturale." In *Opere*, edited by Giorgio Dilemmi, 308–23. Bologna: Commisione per i testi di lingua, 1976.
Ibn al-Nafīs, ʿAlī ibn Abī al-Ḥazm. *The Theologus Autodidacticus of Ibn al-Nafīs*. Translated by Max Meyerhof and Joseph Schacht. Oxford: Clarendon Press, 1968.
Locke, John. *Essay Concerning Human Understanding*. London, 1690.
———. *Some Thoughts Concerning Education*. London, 1693.
Maimonides, Moses. *The Guide for the Perplexed*. Translated by Michael Shwarz. 2 vols. Tel-Aviv, 2002.
Mirandola Giovanni, Pico della. *Oratio de hominis dignitate*. Università degli Studi di Bologna and Brown University, 1999.
———. *Disputationes adversus astrologiam divinatricem*. Edited by Eugenio Garin. 2 vols. Florence, 1946.
———. *Heptaplus*. Translated by Douglas Carmichael. Indianapolis, 1998.
———. "Apologia." In *Opera Omnia*, edited by Gian Francesco Mirandola della Pico. 2 vols. Hildesheim, Ger., 1969.
———. "Commento on Dell'amore celeste e divino." In *Commentary on a Poem of Platonic Love*, translated by Douglas Carmichael. London, 1986.
———. *Syncretism in the West: Pico's 900 Theses (1486)*. Translated by S. A. Farmer. Tempe, Ariz., 1998.
———. *On Being and One: To Angelo Poliziano*. Translated by Paul Miller. Indianapolis, 1998.
Narbonni, Moshe. *Hanhagat Hamitboded*. Bayerische Staatsbibliothek, Munich. Cod. Hebr. 59.
———. "Problems of the Soul and Its Powers." *Daʿat* 23 (1989): 65–88.
———. "Lépitre du libre arbitare de Moise de Narbonne." *Revue des études juives* (1982): 139–67.
———. *Biur Moreh Nevochim*. Bibliothèque Nationale, Paris. Heb. 696.
———. "Beur le-Sefer Moreh Navochim." Vienna: Aus der K. K. Hof und Staatsdruckerei, 1832. Republished in *Sheloshah Kadmone mefarshe ha-Moreh*. Jerusalem, 1960.
———. *The epistle on the possibility of conjunction with the active intellect, by Ibn Rushd; with the commentary of Moses Narbonni; a critical edition and annotated translation by Kalman P. Bland*. New York, 1982.
———. *Sefer Haderushim haTiv'iyot ve-haEloiyot le-Ibn-Rushd*. Bibliothèque Nationale, Paris. Heb. 988.
Pococke, Edward. *A Commentary on the Prophecy of Micah, by Edward Pocock*. London, 1677.
———. *The Book of Common Prayer: Liturgiae Ecclesiae Anglicanae*. Oxford, 1674.
———. *De veritate religionis Christianae: Kitāb fī ṣaḥat al-Sharīʿa' al-Mashīḥyyah*. Translated by Edward Pococke. Oxford, 1660.
———. *The nature of the drink kauhi, or coffe, and the berry of which it is made de-*

scribed by an Arabian phisitian, of Antaki, Da'ud ibn 'Umar. Translated by Edward Pococke. Oxford, 1659.

Rousseau, Jean-Jacques. *Émile, or On Education*. Translated by Barbara Foxley. London, 1911.

Telescope, Tom. *The Newtonian System of Philosophy: Adapted to the Capacities of Young Gentlemen and Ladies, and Familiarized and made Entertaining, by Objects with which they are Intimately acquainted . . . for the Instruction and Rational Entertainment of the Youth of these kingdoms.* London, 1794.

Other Primary Sources

The proximate connection between *Ḥayy Ibn-Yaqẓān* and other major texts was echoed in manifold primary sources that discuss the question of autodidacticism. An interwoven reading of the following selection—collections of letters, contemporary chronicles, autobiographies, travel journals, and other forms of writing—brings to light a web of meanings, values, and practices within which discussions of autodidacticism surfaced and which sparked local controversies regarding mysticism, pedagogy, astrology, and natural philosophy.

Adret, Solomon ben Abraham. *Teshuvot ha-Rashba: Le-Rabenu Shelomoh Aderet; Teshuvot ha-shayakhot la-Miḳra midrash ve-de'ot ve-tsuraf la-hen Sefer Minḥat ḳena'out le-R[aba] Mari de-Lonil*. 5 vols. Jerusalem: Mosad ha-Rav Kuk, 1990.

Avempace. *El régimen del solitario*. Edited and translated by Don Miguel Asín Palacios. Madrid, 1946.

Bacon, Francis. *Sylva sylvarum, or A Naturall History in Ten Centuries*. London, 1626.

Bodley, Thomas. *The life of Sir Thomas Bodley, the honourable founder of the publique library in the University of Oxford, written by himself.* Oxford, 1647.

Boyle, Robert. *Correspondence of Robert Boyle, 1636–1691*. Edited by Michael Hunter, Antonio Clericuzio, and Lawrence M. Principe. 6 vols. London: Pickering and Chatto, 2001.

Campanella, Tommaso. *Civitas solis*. Frankfurt, 1623.

Delmedigo, Elijah. *Beḥinat ha-dat: Examen religionis*. Jerusalem, 1969.

Dhahabī, Muḥammad. *Siyar a'lām al-nubalā'*. 25 vols. Beirut, 1982.

Ficino, Marcilio. *De vita libri tres*. New York, 1978.

Al-Ghazzālī, Abū Ḥāmid Muhammad. *Al-Ma'ārif al-'aqlīyyah wa lubāb al-ḥikmat al-ilahīyya*. Bodleian Library, Oxford University. Pococke 263.

Hakluyt, Richard. *The principal navigations, voyages, traffiques & discoveries of the English nation made by sea or over-land to the remote and farthest distant quarters of the earth at any time within the compasse of these 1600 yeeres*. Vol. 6. New York, 1969.

Ibn Qunfudh, Abu al-'Abbas Ahmad Ibn Husayn. *Al-Fārisīyah fī mabādi' al-dawlah al-ḥafṣīyah: Taqdīm wa-taḥqīq Muḥammad al-Shādhilī al-Nayfar wa-'Abd al-Majīd al-Turkī*. Tunis, 1968.

Ibn Rushd, Muḥammad ibn Aḥmad. *Masāʾil Abī al-Walīd Ibn Rushd (al-Jadd)*. 2 vols. Casablanca, 1992.
Ibn Saʿīd, ʿAlī ibn Mūsá. *Kitāb al-jughrāfīyā*. Beirut, 1970.
Ibn Shahriyār, Buzurg. *ʿAjāib al-Hind: Barruhā wa-baḥruhā wa-jazāʾiruhā; li-Buzurk ibn Shahriyār al-Nākhidhāh al-Ramhurmurzī*; taḥqīq ʿAbd Allāh Muḥammad al-Ḥibshī Abū Ẓaby. United Arab Emirates, 2000.
———. *Livre des merveilles de l'Inde, par Bozorg Fils de Chahriyâr de Râmhormoz; texte Arabe, publié d'après le manuscrit de M. Schefer, collationné sur le manuscrit de Constantinople, par P. A. van der Lith*. Translated by L. Marcel Devic. Leiden, Neth., 1883–86.
———. *The Book of the Marvels of India*. Translated from French into English by Peter Quennell. London, 1928.
Ibn-Tibbon, Shmuel. *Otot ha-shamayim: Samuel Ibn Tibbon's Hebrew Version of Aristotle's Meteorology*. A critical edition with introduction, translation, and index by Resianne Fontaine. Leiden, Neth: Brill, 1995.
Ibn-Tūmart, Muḥammad. *Aʿazz mā yuṭlab wa afḍal mā yuktasab wa anfas mā yudhkhar wa aḥsan mā yuʾmal: Al-ʿilm alladhī jaʾalahu Allāh sabab al-hidāya ilā kulli khayr*. In *Le Livre de Mohammed Ibn Toumert*, by Ignác Goldziher, 1–59. Algiers, 1903.
Grotius, Hugo. *De veritate religionis christianae*. Oxford, 1639.
Locke, John. *The Correspondence of John Locke*. Edited by E. S. De Beer. 8 vols. Oxford, Clarendon Press, 1976–89.
Machiavelli, Niccolò. *History of Florence and of the Affairs of Italy from the Earliest Times to the Death of Lorenzo the Magnificent*. New York, 1960.
Maimonides, Moses. "Igeret Le-Rabbi Shemuel Ibn-Tibon be-ʿInyane tirgum HaMoreh." In *Igrot HaRambam* [Letters of Maimonides]. Jerusalem, 1986.
León, Hebreo. *Dialoghi d'amore, a cura di Santino Caramella*. Bari, 1929.
Leoniceno, Niccolò. *In libros Galeni e Greca in Latinam linguam a se translatos prefatio communis*. Venice, 1508.
Manfredi, Girolamo. *Libre de homine*. Bologna, 1474.
Manuzio, Aldo. *Strozii poetae pater et filius*. Venice, 1515.
Mari, Aba. *Sefer minhat kanaut*. In vol. 5 of *Teshuvot ha-Rashba: Le-Rabenu Shelomoh Aderet; Teshuvot ha-shayakhot la-Miḳra midrash ve-deʿot ve-tsuraf la-hen Sefer Minḥat ḳenaʾout le-R[aba] Mari de-Lonil*, by Solomon ben Abraham Adret. Jerusalem: Mosad ha-Rav Kuk, 1990.
Al-Masʿūdī, Abu al-Hassan. *Murūj al-dhahab wa-maʿādain al-jawhar*. 4 vols. Beirut, 1965.
Mirandola della, Gianfrancesco Pico. "The Life of Pico," translated by Thomas More. In *The Yale Edition of the Complete Works of Thomas More*, 1:47–75. New Haven: Yale University Press, 1997.
More, Thomas. "De tristitia Chrisiti." In volume 14 of *Complete Works of St. Thomas More*, edited by Clarence H. Miller. New Haven, Conn.: Yale University Press, 1963–ca. 1997.
———. *Here is cōteyned the lyfe of Johan Picus Erle of Myrandula, a grete lorde of Italy,*

an excellent connynge man in all sciences [beta] verteous of lyuynge. With dyuers epystles [beta] other werkes of [the] sayd Johan Picus. London, 1510.

———. De optimo reip. statv, deque noua insula Vtopia: Libellus uere aureus, nec minus salutaris quàm festiuus. Leuven, Belg., 1516.

Philosophical Transactions of the Royal Society 6, no. 73 (1671): 2214.

Pliny, the Elder. Natural History. Translated by H. Rackham. 10 vols. Cambridge, Mass.: Harvard University Press, 1967–75.

Poliziano, Angelo. Letters. Edited and translated by Shane Butler. Vol. 1. Cambridge, Mass.: I Tatti Renaissance Library, Harvard University Press, 2006.

Al-Qaṭṭān, Ḥasan Ibn ʿAlī al-Marrākushī. Nuẓum al-jumān li-tartīb mā salafa min akhbār al-zamān. Beirut, 1990.

Robson, Charles. Nevves from Aleppo A letter written to T.V. B. of D. Vicar of Cockfield in Southsex· By Charles Robson Master of Artes, fellow of Qu: Col: in Oxford, and preacher to the Company of our English Merchants at Aleppo. Containing many remarkeable occurrences obserued by him in his iourney, thither. London, 1628.

Savonarola, Girolamo. Selected Writings of Girolamo Savonarola: Religion and Politics, 1490–1498. Edited by Donald Beebe, Alison Brown, and Giuseppe Mazzotta. New Haven, Conn.: Yale University Press, 2006.

Suhrawardī, Yaḥyā ibn Ḥabash. Risālāt-i "Ḥayy ibn Yaqẓān." Translated by Ḥamdī Sanadājī and Burhān al-Dīn. Teheran, 1977.

Trimmer, Sarah. An Easy Introduction to the Knowledge of Nature, and Reading the Holy Scriptures. Adapted to the Capacities of Children. London, 1799.

Wilkins, John. An Essay Towards a real Character and a Philosophical Language. London, 1668.

Wood, Anthony. Athenæ Oxonienses: An exact history of all the writers and bishops who have had their education in the most ancient and famous University of Oxford. 2 vols. Oxford, 1691.

Secondary Literature

Since the late nineteenth century the story of Ḥayy Ibn-Yaqẓān has employed scholars of various fields, from literature, philosophy, and religious studies to history and the history of science. The culturally rich list of selected items below exemplifies the range of disciplinary field of interest and the variety of languages that took up the topic and dealt with it through diverse intellectual and cultural angles.

MEDIEVAL ISLAMIC PHILOSOPHY AND HISTORY

Since the story of Ḥayy Ibn-Yaqẓān is Ibn-Tufayl's sole extant philosophical treatise, most scholarship discusses his philosophy through analysis and comparison of the story of Ḥayy with other contemporary philosophical works. Ḥayy Ibn-Yaqẓān comes up as a turning point in medieval philosophy in general and in Islamic philosophy in particular. It represents the philosophical synthesis of the Eastern and the Western Islamic traditions. Previous scholarship, however, ignored the contextual connection with the reception of Sufism in Andalusia that took place at the same time that Ibn-

Tufayl composed his story. It is my argument in chapter 1 that such texts should be read against accounts of the political and cultural contexts in Andalusia. The selected items below, therefore, include works that focus on certain aspects of the story as well as works on the history of Islamic Iberia, most particularly regarding the reception of Sufism in Andalusia.

Amīn, Aḥmad. "*Ḥayy ibn Yaqẓān*": *li-Ibn Sīnā wa-Ibn Ṭufayl wa-Suhrawardī*. Damascus: Dār al-Madá lil-Thaqāfah wa-al-Nashr, 2005.
Bürgel, J. C. "Ibn-Tufayl and His Hayy Ibn-Yaqzan: A Turning Point in Arabic Philosophical Writing." In *The Legacy of Muslim Spain*, edited by Salma Khadra Jayyusi and Manuela Marín, 2:830–49. Leiden, Neth.: Brill, 1992.
Colville, Jim, trans. *Two Andalusian Philosophers*. London: Kegan Paul International, 1999.
Conrad, Lawrence. *The World of Ibn-Tufayl: Interdisciplinary Perspectives of "Hayy Ibn-Yaqzan."* Leiden, Neth.: Brill, 1996.
Dozy, Reinhart Pieter Anne. *Historia de los musulmanes de España hasta la conquista de los Almoravides*. 4 vols. Madrid, 1930.
———, ed. *The History of the Almohades, by 'Abdu al-Wāhid al-Marrākushī*. Leiden, Neth.: 1881.
Elmusa, Sharif S., ed. *Culture and the Natural Environment: Ancient and Modern Middle Eastern Texts*. Cairo: American University in Cairo Press, 2005.
Fierro, Maribel. "La religion." In *Historia de España*, vol. 8, *Los Reinos de Taifas: Al-Andalus en el siglo XI*, edited by Ramón Menéndez Pidal. Madrid, 1994.
———. "The Polemic about the Karāmāt Al-awliyā' and the Development of Ṣufism in al-Andalus (Fourth / Tenth–Fifth / Eleventh Centuries)." *Bulletin of the School of Oriental and African Studies* 55 (1992): 236–49.
Fletcher, Madeleine. "The Almohad Tawḥīd: Theology Which Relies on Logic." *Numen* 38, no. 1 (1991): 110–27.
Garden, Kenneth. "Al-Ghazālī's Contested Revival: Iḥyā' 'Ulūm al-Dīn and Its Critics in Khorasan and the Maghrib." Ph.D. diss., University of Chicago, 2005.
Goodman, Evan Lenn, trans. *Ibn Tufayl's "Hayy ibn Yaqzān": A Philosophical Tale*. New York, 1972.
Gutas, Dimitri. "Ibn-Tufayl on Ibn-Sina's Eastern Philosophy." *Oriens* 34 (1994): 222–41.
Hawi, Sami S. *Islamic Naturalism and Mysticism: A Philosophic Study of Ibn Ṭufayl's "Ḥayy bin Yaqṣān."* Leiden, Neth.: Brill, 1974.
Hopkins, J. F. P. "The Almohade Hierarchy." *Bulletin of the School of Oriental Studies* 16, no. 1 (1954): 93–112.
el Hour, Rachid. "The Andalusian Qaḍī in the Almoravid Period: Political and Judicial Authority." *Studia Islamica* 90 (2000): 67–83.
Mediano, Fernando Rodríguez. "Biografias Almohades en el Taṣawwuf de al-Tālidī." *Estudios Onomástico-Biográficos de al-Andalus* 10 (2000): 167–92.
Wat geen oog heeft gezien, geen oor heeft gehoord en in geen mensenhart is opgekomen: De geschiedenis van Hayy ibn Yaqzan / Abu Bakr Muhammad ibn Tufayl;

Uit het Arabisch vertaald en ingeleid door Remke Kruk. Amsterdam: Meulenhoff, ca. 1985.

Serrano de Haro, Agustín. *Abentofáil y el filósofo autodidacto.* Gaudix, Spain, 1926.

LATE MEDIEVAL JEWISH HISTORY AND PHILOSOPHY

Iberian and Provençal Jewish culture played a key role in further circulating the text to other cultural contexts beyond Arabic or Islamic boundaries. Beyond the plain philosophical text of *Ḥayy Ibn-Yaqẓān,* here we find commentaries on the text that bring out richer context. The selected items below represent not only accounts of the Hebrew text of the story but also extensive correspondence and notarial literature that invokes a new web of meanings concerning pedagogy and autodidacticism within which the story of *Ḥayy* was embedded.

Abrahams, Israel, ed. *Tsavaʾot geʾone Yisrael.* 2 vols. Philadelphia, 1926.

Adret, Solomon ben Abraham. *Teshuvot ha-Rashba: Le-Rabenu Shelomoh Aderet; Teshuvot ha-shayakhot la-Mikra midrash ye-deʾot ye-tsuraf la-hen Sefer Minḥat ḳenaʾout le-R[aba] Mari de-Lonil.* 5 vols. Jerusalem: Mosad ha-Rav Kuk, 1990.

Emery, Richard Wilder. *The Jews of Perpignan in the Thirteenth Century: An Economic Study Based on Notarial Records.* New York: Columbia University Press, 1959.

Goitein, S. D. "From the Mediterranean to India: Documents on the Trade to India, South Arabia, and East Africa from the Eleventh and Twelfth Centuries." *Speculum: A Journal of Medieval Studies* 29, vol. 2 (1954): 181–97.

Halbertal, Moshe. *Between Torah and Wisdom: Rabbi Menachem ha-Meiri and the Maimonidean Halakhists in Provence* [Ben Torah Le-Hochma: Rabbi Menachem Hamairi u Baʾali Hahalach Ha-Maimonim be Provence]. Jerusalem: Hebrew University Magnes Press, 2000.

Hayoun, Maurice. "Lépitre du libre arbitare de Moise de Narbonne." *Revue des études juives* (1982): 139–67.

———. "Moshe Narbonni: Beʾayot haNefesh u-Kohoteha" [Moses Narbonni: Problems of the soul and its powers]. *Daʾat* 23 (1989): 65–88.

Holzman, Gitit. "Torat ha-nefesh veha-sekhel be-haguto shel R. Mosheh Narbonni : Al pi beurav le-khitve Ibn Rushd, Ibn Tufil, Ibn Bagʾah ve-Algazali." Ph.D. diss., Ha-Universitah ha-Ivrit, Jerusalem, 1996.

Idel, Moshe. *Studies in Ecstatic Kabbalah.* New York, 1988.

———. "The Ladder of Ascension: The Reverberations of a Medieval Motif in the Renaissance." In *Studies in Medieval Jewish History and Literature,* edited by Isadore Twersky, 2:83–93. Cambridge, Mass.: Harvard University Press, 1984.

———. "The Anthropology of Yohanan Alemanno: Sources and Influences." *Topoi* 7 (1988): 201–10.

Ivry, Alfred Lyon. "Moses of Narbonne's Treatise, *The Perfection of the Soul:* A Partial Edition from the Paris MS." Ph.D. diss., Brandeis University, 1963.

Loeb, Isidore. "Liste nominative des Juifs de Barcelone." *Revue des études juives* (1882): 57–77.

Mari, Aba. *Sefer Minhat Kanaut*. In vol. 5 of *Teshuvot ha-Rashba: Le-Rabenu Shelomoh Aderet; Teshuvot ha-shayakhot la-Mikra midrash ye-de'ot ye-tsuraf la-hen Sefer Minḥat ḳena'out le-R[aba] Mari de-Lonil*, by Solomon ben Abraham Adret. Jerusalem: Mosad ha-Rav Kuk, 1990.

Miller, Larry. "Philosophical Autobiography: Moshe Narbonni's Introduction to his Commentary on *Ḥayy Ibn-Yaqẓān*." In *The World of Ibn Tufayl*, edited by Lawrence Conrad, 229–37. Leiden, Neth.: Brill, 1996.

Pick, Shlomo H. "The Jewish Communities of Provence before the Expulsion in 1306." Ph.D. diss., Bar-Ilan University, 1996.

Rosenthal, E. I. "The Place of Politics in the Philosophy of Ibn Bajja." *Islamic Culture* 25 (1951): 187–211.

———. "Political Ideas in Moshe Narbonni's Commentary on Ibn Tufail's Hay B. Yaqzan." In *Hommage à Georges Vajda*, edited by Gérard Nahon and Charles Touati, 227–34. Leuven, Belg., 1980.

Sirat, Colette. "Perkei Moshe le-Moshe Narbonni" [Moshe's chapters of Moshe Narbonni]. *Tarbiz: A Quarterly for Jewish Studies* 39, no. 3 (1970): 287–306.

Vajda, Georges. "Comment le philosophe Juif Moise de Narbonne, commentateur d'Ibn Tufayl, comprenait-il les paroles extantiques (satahāt) de Soufis?" In *Mélanges Georges Vajda: Études de pensée, de philosophie et de littérature juives et arabes; In memoriam*, edited by G. E. Weil, 275–81. Hildesheim, Ger.: 1982.

Vidal, Pierre. "Les Juifs de ancien comtés Roussilon et de Cerdagne." *Revue des études juives* 16 (1888): 170–87.

RENAISSANCE CULTURE

Through the movements of Hebraism and Christian Kabbalah the Hebrew commentaries on *Ḥayy Ibn-Yaqẓān* were introduced to Renaissance thinkers. Pico della Mirandola's Latin translation of the story is plain and gives no indication regarding its context. Pico's interest in the story, however, comes to light when we read a variety of contemporary texts against the structure of the story of *Ḥayy*. The selected items below include articles and works that describe the cultural crisis in late fifteenth-century Florence and others that introduce micro details of the Pico's relations with his colleagues during that crisis. These works expand on the meaning of autodidacticism as a resolution of contemporary controversies regarding predestination and self-molding.

Altrocchi, Rudolph. "Dante and Tufail." *Italica* 15, no. 3 (1938): 125–28.

Bacchelli, Franco. "Pico della Mirandola traduttore di Ibn Tufayl." *Giornale critico della filosofia Italiana* 6, no. 13 (1993): 1–25.

———. *Giovanni Pico e Pier Leone da Spoleto: Tra filosofia dell'amore e tradizione cabalistica*. Florence, 2001.

Branca, Vittore. "Alla ricerca di libri e di docenti nel Vento umanistico." In *Poliziano: E l'umanesimo della parola*, 134–55. Turin: Einaudi, 1983.

Castelli, Patrizia. "L'oroscope di Pico." In *Pico, Poliziano e l'Umanesimo di fine quattrocento*, edited by Paolo Viti, 225–29. Florence: Olschki, 1994.

Copenhaver, Brian. "The Secret of Pico's Oration." *Midwest Studies in Philosophy* 26 (2002): 56–81.

Ficino, Marcilio. *De vita libri tres*. New York, 1978.

Fioravanti, Gianfranco. "Pico e l'ambiente ferrarese." In *Giovanni Pico della Mirandola: Convegno internazionale di studi nel cinquecentesimo anniversario della morte (1494–1994)*. Edited by Gian Carlo Garfagnini, 1:157–72. Florence, 1997.

Frati, Ludovico. "La morte di Lorenzo de' Medici e il suicidio di Pier Leoni." *Archivio storico Italiano* 5, no. 4 (1889): 255–60.

Fregoso, Antonio Fileremo. "De lo istinto naturale." In *Opere*, edited by Giorgio Dilemmi, 308–23. Bologna: Commisione per i testi di lingua, 1976.

Grafton, Anthony. "Giovanni Pico della Mirandola: Trials and Triumphs of an Omnivore." In *Commerce with the Classics: Ancient Books and Their Renaissance Readers*. Ann Arbor, 1997.

Guerra-Coppioli, Luigi. "Pierleone da Spoleto, medico e filosofo: Note biografiche con documenti inediti." *Bollettino della R. Deputazione di storia patria per l'Umbria* 23 (1915): 387–430.

Heering, J. P. *Hugo Grotius as Apologist for the Christian Religion: A Study of His Work "De veritate religionis christianae" (1640)*. Leiden, Neth.: Brill, 2004.

Lelli, Fabrizio, ed. "*Hay Ha-'Olamim*" (*L'Immortale*) [by] Yohanan Alemanno. Florence, 1995.

León, Hebreo. *Dialoghi d'amore, a cura di Santino Caramella*. Bari, 1929.

Leoniceno, Niccolò. *In libros Galeni e Greca in Latinam linguam a se translatos prefatio communis*. Venice, 1508.

Lesley, Arthur. "The Song of Solomon's Ascents by Yohanan Alemanno: Love and Human Perfection According to a Jewish Colleague of Giovanni Pico della Mirandola." Ph.D. diss., University of California, Berkeley, 1976.

Moody, Ernst. "Galileo and Avempace: The Dynamics of the Leaning Tower Experiment," pt. 1, *Journal of the History of Ideas* 12, no. 2 (1951): 163–93.

———. "Galileo and Avempace: The Dynamics of the Leaning Tower Experiment," pt. 2, *Journal of the History of Ideas* 12, no. 3 (1951): 375–422.

———. "Empiricism and Metaphysics in Medieval Philosophy." *Philosophical Review* 67, no. 2 (1958): 145–63.

Di Napoli, Giovanni. *Giovanni Pico della Mirandola e la problematica dottrinale del suo tempo*. Rome, 1965.

Novak, B. C. "Giovanni Pico della Mirandola and Jochanan Alemanno." *Journal of the Warburg and Courtauld Institutes* 45 (1982): 125–47.

Radetti, Giorgio. "Un'aggiunta alla biblioteca di Pierleone Leoni da Spoleto." In *Rinascimento: Rivista dell'Istituto Nazionale di Studi Sul Rinascimento*, 5:87–99. Florence, 1965.

Rutkin, Darrell. "Astrology, Natural Philosophy and the History of Science, ca. 1250–1700: Studies Toward an Interpretation of Giovanni Pico della Mirandola's Disputationes Adversus Astrologiam Divinatricem." Ph.D. diss., Indiana University, 2002.

———. "Astrological Conditioning of Same-Sexual Relations." In *The Sciences of Homosexuality in Early Modern Europe*, edited by Kenneth Borris and George Rousseau, 165–201. London, 2008.

Toomer, Gerald J. *Eastern Wisdom and Learning: The Study of Arabic in Seventeenth-Century England*. Oxford: Oxford University Press, 1996.

Viti, Paolo, ed. *Pico, Poliziano e l'Umanesimo di fine Quattrocento*. Florence: Olschki, 1994.

Wirszubski, Chaim. *Pico della Mirandola's Encounter with Jewish Mysticism*. Cambridge, Mass.: Harvard University Press, 1989.

ENLIGHTENMENT

The reception of Edward Pococke's *Philosophus autodidactus* generated great interest during the Enlightenment and stimulated scholars, mostly in comparative literature, who looked for intertexualites between Enlightenment philosophy and culture and the story of Ḥayy. The selected items show the great scholarly interest in the connection between popular heroes like Robinson Crusoe, philosophers like Locke, and the rise of children's literature.

'Abbās, Ḥasan Maḥmūd. "*Ḥayy ibn Yaqẓān*" *wa-Rūbinsūn Kirūzū: Dirāsah muqāranah*. al-Ṭabʿah 1. Beirut: Al-Muʾassasah al-ʿArabīyah lil-Dirāsāt wa-al-Nashr, 1983.

———. *Der Ur-Robinson, mit e. Nachw. von Otto F. Best. Ins Dt.* Translated by Otto F. Best. Munich: Matthes u. Seitz, 1987.

———. *Hajj ibn Jaqzan der Naturmensch: Ein philosophischer Robinson-Roman aus dem arabischen Mittelalter*. Leipzig: Kiepenheuer, 1983.

Attar, Samar. *The Vital Roots of European Enlightenment: Ibn-Tufayl's Influence on Modern Thought*. Lanham, Md., 2007.

Ekhtiar, Shelly. "Ḥayy ibn Yaqẓān: The Eighteenth-Century Reception of an Oriental Self-Taught Philosopher." *Studies on Voltaire and the Eighteenth Century*, no. 302 (1992): 217–45.

Kruk, Remke. "An 18th-Century Descendant of Ḥayy ibn Yaqẓān and Robinson Crusoe: Don Antonio de Trezzanio." *Arabica* 34, no. 3 (1987): 357–65.

Pastor, Antonio. *The Idea of Robinson Crusoe*. Watford, U.K., 1930.

Pickering, Samuel. *John Locke and Children's Books in Eighteenth Century England*. Knoxville: University of Tennessee Press, 1981.

Russell, G. A. "The Impact of *Philosophus Autodidactus*: Pocockes, John Locke, and the Society of Friends." In *The "Arabick" Interest of the Natural Philosophers in Seventeenth-Century England*, 224–65. New York: Brill, 1994.

Zaidan, Samira H. *A Comparative Study of Haiu Bnu Yakdhan, Mowgli and Tarzan*. Great Missenden, U.K.: Red Squirrel Books, 1998.

INDEX

Abbās, Ḥasan Maḥmud, 162n6
Abramo, Isaac, 76
Abubachar. *See* Ibn-Tufayl, Abū Bakr Muhammad
Abu Bakr al-Hasan ibn al-Khasib (Albubather, Abubater), 71, 153n15
Abul'afia, Rabbi Meir HaLevi, 146n1
Abulaifia, Avraham, 150n44
Abū Ya'qūb Yūsuf, 16, 25, 27, 28, 29–30
Academia Secretorum Naturae (scientific society), 131, 134
An Account of Philaretus During His Minority (Boyle), 115
An Account of the Oriental Philosophy (trans. *Ḥayy Ibn-Yaqẓān*; Keith), 10, 119–20
Aims of Philosophers (al-Ghazzālī), 60
Albertus Magnus, 156n35
Alemanno, Jochanan, 64, 73–75, 86, 91, 130, 154n21, 155n27
Aleppo, 15, 101, 106, 107, 108, 113, 163n10
Alfonso VII (king of Castile), 29
'Ali b. Yūsuf b. Tāshufin, 16
Almohad dynasty, 16, 21, 25–31, 40, 46
Almoravid dynasty, 16, 19, 21, 25–26, 30, 142n20
Altamira Cave (Spain), 39
Altrocchi, Rudolph, 140n13
Amores mei (My Loves; Pico della Mirandola), 88
Antaki, Da'ud ibn 'Umar, 111
Antūn, Farah, 41
Apologia (Pico della Mirandola), 68, 77
Arabic language, 10, 27, 44, 71, 101, 104, 109, 110; sources in, 41, 107–8, 149n15; translations to and from, 9–10, 42, 48, 49, 111–14, 120–21, 129
Aristotle, vii, 30, 63, 87, 89, 134, 144n40; on climate, 34; and spontaneous generation, 118; *vs.* traditional authority, 51, 52
Ashwell, George, 10
"The Aspiring Naturalist" (story; Boyle), 115
astrology, 13, 156n37; and death of Lorenzo de' Medici, 78–80; in Florence, 70, 129, 130; and homosexuality, 84, 86; Pico on, 65, 66, 71, 76–78, 88, 90, 91, 92, 95, 97, 152n11, 155n30; predictions of, 130, 156n31, 158n65
Astronomicon (Manilius), 159n74
astronomy, 22, 26, 53, 58, 105, 109, 120; *vs.* astrology, 71; and climate, 31–32; and Ḥayy, 3, 23–24
Attar, Samar, 141n13, 162n6
authority, 6, 43, 81, 102, 129, 142n9; *vs.* autodidacticism, 49, 59, 138; and Boyle, 115–16; challenges to, 15, 50–51, 62, 137; from consensus, 112–14; *vs.* experience, 15, 18, 19, 25, 40, 46, 120, 125–30; *vs.* oral tradition, 15, 19; *vs.* philosophy, 51–53; Pico *vs.*, 66, 68, 71, 153n11; and Sufism, 15, 16, 19, 25, 30, 144n39
autodidacticism: and child prodigies, 66, 89, 115; in children's literature, 5–6; and the Enlightenment, 9, 69, 126; and experimentalism, 13, 126, 138; history of, 126–38; Horace on, vii, 68, 100, 102, 126, 137; and Kabbalah, 150n44; and modernity, 1, 138; and nature, 59, 126; and solitude, 30, 49, 62; and sudden perception, 59–61, 150n44; and Sufism, 15–17, 19; and theories of knowledge, 102; *vs.* traditional authority, 59, 138; and utopianism, 126–28; and wild boys, 1–2, 4, 9, 41
Avempace (Ibn-Bājja), 30, 46, 58, 116, 144n40, 148nn6, 11, 150n38, 167n44

Index

Averroes (Abū 'l-Walīd Muḥammad Ibn-Rushd), 10, 41, 51, 129, 144n40, 154n17, 167n44; and Abū Ya'qūb Yūsuf, 29–30; and grandfather, 29, 144n35; and Ibn-Tufayl, 28, 144n35; and Narbonni, 44, 60, 147nn5, 6
Avicenna, 31, 140n13, 141n3

Bacchelli, Franco, 76, 160n100
Bacon, Francis, vii, 99–100, 115, 127, 135–36, 137
Barbaro, Ermolao, 158n61
Barret, William, 162n10
Bar-Shmuel, Moshe, 54
Bartolomeo, Fra, 94
Baṭalyawsī, 155n27
Baxter, Richard, 167n42
Bellanti, Lucio, 156n31
Bellashom, Bonjuhes, 148n8
Ben-Aderet, Shlomo, 50–55, 58, 61, 62, 150n31, 151n44
Ben-Avuyya, Elisha, vii, 121
Ben-Israel, Menashe, 63, 164n24
Benivieni, Domenico, 85
Benivieni, Girolamo, 65, 76, 84, 87, 89, 156n31
Ben-Yehiel, Asher, 46, 61–62
Berlin, Isaiah, 127
Biagioli, Mario, 99
Bible, 50, 73, 86, 107–8, 109, 127, 170n6. See also Moses; Song of Songs
al-Bīrūnī, Abū Rayḥān, 34
blank slate (tabula rasa), vii, 1–2, 5, 9, 13, 36, 60, 120–24
Bochart, Samuel, 164n19
Bodley, Thomas, 163n10
Bolognini, Giovanni, 160n100
Book of Curiosities (Kitāb gharā'ib al-funūn wa-mulaḥ al-al-uyūn), 33
Book of Geography (Kitāb al-jughrāfiyā), 36
Book of Wonders of India (Kitāb 'ajā ib al-hind; Buzurg ibn-Shahriyār), 32, 35–36, 37
The Book on the Shape of the Earth (Kitāb ṣūrat al-arḍ; al-Khwārizmī), 34
Botticelli, Sandro, 65
Bottonio, Fra Timoteo, 96
Boyle, Robert, 101, 108, 111–20, 164n26, 166nn39,41; and Pococke, 111, 112–16, 163n14, 164n27
Burckhardt, Jacob, 152n11
Burgos, Abner de, 56
Buzurg ibn-Shahriyār, 32

Calderini, Antonio, 93
Campanella, Tommaso, 99, 127, 132–34, 135
Cassirer, Ernst, 152n11
Catastini, Angelo, 156n31
caves, 24, 38–40
Cescres, Hasdai, 63
Cesi, Federico, 99
Charles I (king of England), 110
Charles III (king of England), 123
Charles VIII (king of France), 68, 93
children, 131, 132, 134; literature for, 5–6, 8, 138; Locke on, 121–22; and nature, 4–9; as prodigies, 4, 5, 9, 13, 40, 66, 89, 115, 138; wild, 1–2, 4, 6, 9, 138. See also blank slate
The Children's Miscellany (Day), 6, 8
Christianity, 127, 129; censorship by, 65, 67, 78, 131, 134; conversions to, 43, 56, 76, 112, 113, 154n18; and experimentalism, 118–20; and Jews, 42–43. See also Bible; God; Savonarola, Girolamo
The Christian Virtuoso (Boyle), 118–19, 120
City of the Sun (Civitas solis; Campanella), 127, 132–34, 135
Civil War (England), 110, 112, 123, 137, 163n10
Clark, Edward, 123
climate, 31–32, 34–35, 91–92, 95–96, 118
coffee, 111, 164n26
The Collapse of Philosophers (Averroes), 28
The Complete Treatise on the Prophet's Biography (al-Risāla al-Kāmilīyah fī al-sīrah al-nabawīyyah; Ibn al-Nafīs), 41
Comtiano, Mordechai, 110
Conclusiones philosophicae, cabalasticae et theologicae (Theses; Pico della Mirandola), 67, 76, 77
The Confirmation of the Rule (painting; Ghirlandaio), 81
Conrad, Lawrence, 140n13
Constantinople, 106, 109–10, 111, 163n10
Constanzo, Jacopo, 160n90
contemplation: vs. active life, 48–49, 57, 58, 90, 127, 129, 132; embodied (dhikr), 19; vs. experience, 5, 24, 58, 65, 119, 128; of God, 57, 70
Copenhaver, Brian, 153n11
Copernicus, Nicolaus, 63
Cornell, Vincent, 27
cross-cultural exchange, 31, 34, 105–6, 124, 138, 162n10
Cross-Cultural Scientific Exchanges in the Eastern Mediterranean (Ben-Zaken), 13

Culpeper, Nicholas, 167n50
culture: and Boyle, 111, 116, 119; of Catalonia vs. Provence, 42–43, 44, 46, 49, 54, 61, 62; and climatic zones, 34–35, 92, 95–96; Eastern, 23, 31, 33–34, 92; European, 127–28, 161n6; Florentine, 65, 66, 70, 81, 82, 83, 93, 95, 96, 130; and Ḥayy, 12–14, 15, 25, 38, 40, 101, 128–29, 130, 138, 140n13, 141n3, 145n56; intellectual, 27, 57, 62, 105, 124, 125, 161n6; and knowledge, 9, 102; modern, 41, 138; of Near East, 111, 120; oral, 15, 19; popular, 1–2, 21, 78, 102, 138; and public speeches, 50–51; Savonarola on, 95–96

Dallam, Thomas, 106
Dante Alighieri, 140n13
Dapper, Olfert, 107
Darwish, Ahmad, 108
David (statue; Donatello), 80–81
Day, Thomas, 6, 8, 9
De ente et uno (On Being and the One; Pico della Mirandola), 86, 89
Defoe, Daniel, 1, 9
Dei, Bartolomeo, 80
Delaroche, Paul, 68, 69
Dell'amore celeste e divino (Benivieni), 84, 156n31
della Porta, Giambattista, 131, 134
Delmedigo, Elijah, 71, 154n17, 155n28
Delmedigo, Joseph, 63, 151n51
De lo istinto natural (poem; Fregoso), 97–99, 131
De miraculis occultis naturae (Lemnius), 167n50
De tristitia Christi (More), 102
De veritate religionis Christianae (Grotius), 111–14, 166n42
De vita libri tres (Ficino), 90
Dialoghi d'amore (Ebreo), 86
Dialogue on Adam and Eve (Isotta Norarola), 152n4
direct sensible intuition (*mushāhadah*), 21–22, 23, 39, 59–60
Disputationes adversus astrologiam divinatricem (Pico della Mirandola), 65, 71, 77–78, 80, 83, 84, 86, 89–93, 96–97, 152n11
Dozy, Reinhard, 17

Easy Introduction to the Knowledge of Nature (Trimmer), 9
Ebreo, Leone, 84, 86
Eichhorn, Johann Gottfried, 10
Ekhtiar, Shelly, 162n6

Elizabeth I (queen of England), 105–6
Émile (Rousseau), 6, 7, 137–38
empiricism, 5, 9, 13; and Locke, 120–24, 137; and Pococke, 101, 104, 124, 137
England, 101–25; Civil War in, 110, 112, 123, 137, 163n10; Jews in, 164n24; and Ottoman Empire, 105–6; and trade with Near East, 105–6, 107, 124, 162n10, 165n34; utopianism in, 131–33
the Enlightenment, 4–5, 13, 123–24, 140n13, 161n6; autodidacticism in, 9, 69, 126
Epicurus, 118, 167n47
Epistle on Free Will (Mamar be behira; Narbonni), 56
Erasmus of Rotterdam, 131
Essay Concerning Human Understanding (Locke), vii, 5, 121–23, 124, 137, 161n6
Essay on Spontaneous Generation (Boyle), 116–17
Europe: culture of, 127–28, 161n6; experimentalism in, 127, 141n13; and medieval Islam, 140n13
Evelyn, John, 165n39
experience, 129–37, 168n58; vs. authority, 15, 18, 19, 25, 40, 46, 100, 120, 125–30; and autodidacticism, 126; vs. contemplation, 5, 24, 58, 65, 119, 128; and education, 5, 59; and knowledge, 19, 21–22; Locke on, 121–22; Pico on, 65, 68, 71, 95
experimentalism, 5, 30, 61, 110, 120, 129, 156n35; and autodidacticism, 126, 138; and Bacon, 99–100, 135; and Christianity, 118–20; and dissection, 3, 11, 12, 59, 98–99; in Europe, 127, 141n13; and Ḥayy, 13, 128; and Pico, 71, 77, 99; and Pococke, 101, 102–5, 111, 124, 137, 161n6; and reliable testimonies, 112–14; and spontaneous generation, 116, 124, 125

Fables of Bidpai (*Panchatantra*), 4, 9, 139n7
Fatahallah, Shaykh, 108
Ficino, Marcilio, 65, 87, 89, 90, 156n37
Fletcher, Madeleine, 143n26
A Florentine Diary (Landucci), 96
free will, 56, 75, 95, 97, 153n11
Fregoso, Antonio, 97–99, 130–31, 160n100

Galileo Galilei, 63, 116, 134
Garden, Kenneth, 17
A Gate of Desire (Sha'ar ha-ḥeshek; Alemanno), 73
Gauthier, Léon, 141nn1,3
Geographia (Ptolemy), 34–35

al-Ghazzālī, Abū Ḥāmid, 15, 29, 30, 60, 129, 141n3, 144n35, 147n6; and gazelle, 22, 38; and Ibn-Tufayl, 21–23; and Ibn-Tūmart, 25; and Sufis, 17, 19–23, 25–26, 28; theology of, 16–19

Ghirlandaio, Domenico, 65, 81

God: and astrology, 80; and autodidacticism, 128, 132, 135; and blank slate, 60; Boyle on, 119; coupling with, 66, 70, 73, 75, 86; existence of, 112, 168n56; al-Ghazzālī on, 18, 19, 23; and Ḥayy, 2, 4, 23, 24, 74, 89, 141n3; Ibn-Tufayl on, 22, 24; idea of, 121–23; knowledge of, 21, 22, 24, 28, 105, 111, 119, 125, 141n3; Locke on, 121–23; and magic, 76; Maimonides on, 48; names of, 19, 155n23; Narbonni on, 44, 46, 57, 58, 60, 61, 62, 147n5, 151n46; oneness of, 10, 26, 29, 30, 142n8; and Savonarola, 93, 95, 97; in Sufism, 15, 19, 21, 22, 40; union with, 18, 58, 87, 153n11

Grafton, Anthony, 92, 153n11

Greaves, John, 109–10, 123, 163n12

Greaves, Thomas, 108, 109

Greek Orthodox Church, 109, 110

Greeks, 49, 55, 140n13; and Pico, 76, 92; and utopias, 126–27, 128

Grotius, Hugo, 110, 111–14, 165n39

Guadix (town), 16, 17, 36; caves in, 38, 39

Guarico, Luca, 156n31

Guicciardini, Niccolò di Braccio, 83

Guicciardini, Piero, 83

Guide for the Perplexed (Maimonides), 44–46, 48–49, 56, 63, 121, 146n1, 148nn6,11, 150n42

Guintini, Francesco, 156n31

Gutas, Dimitri, 142n3

Halfon ben Nethanel, 35

ha-Meiri, Menachem, 54, 58

HaNasi, Sheshet, 146n1

Harborne, William, 105–6

Harlay de Cèsy, Philippe, 163n12

Hartlib, Samuel, 114–15, 164n26

Ha-Salmi, Shmuel, 150n31

Ḥayy ha-'olamim (Alemanno), 73

Ḥayy Ibn-Yaqẓān (Alive Son of the Vigilant; Ibn-Tufayl): extant versions of, 141n1; and Far East, 40; history of, 128–37; and pedagogical controversy, 52, 57, 59, 61; Pico's translation of, 71–76, 88–89, 97; Pococke's copy of, 20; story of, 2–4; title of, 57; translations of, 9–10, 11, 12, 13, 15, 42, 63, 101, 119, 129, 146n1

Ḥayy Ibn-Yaqẓān (character), 2–4, 8; awareness of, 38–39; as cartoon, 41; and hierarchy of knowledge, 23–25; and mother gazelle, 2–3, 11, 12, 22, 38, 55, 59

Hayyim, Orah, 148n7

Hebrew language, 42, 48, 71, 101, 120, 129, 146n1, 154n17

Heptaplus (Pico della Mirandola), 73, 75, 77, 130

Het leven van Hai ebn Yokdan (The life of Ḥayy Ibn-Yaqẓān; trans. S.D.B.), 10, 11

Hieroglyphica (Valeriano), 93

Hipparchus, 35

History and Demonstrations Concerning Sunspots and their Phenomena (Galileo), 134

History of Florence (Machiavelli), 68, 80

The History of Hai Eb'n Yockdan an Indian Prince, or The Self-Taught Philosopher (trans. Ḥayy Ibn-Yaqẓān; Ashwell), 10

The History of the Almohads (al-Marrākushī), 25

History of the Dynasties of Abu'l-Faraj (Pococke), 111

Hobbes, Thomas, 101

Holten, Albrecht von, 10

Holtzman, Gitit, 147n5

Homer, 126

homosexuality, 65, 70, 80–83, 92, 93, 129, 130; and Pico, 81, 84, 86–89, 158n61; repression of, 83, 158n59; and Savonarola, 95, 96

Honorius III, Pope, 82

Horace, vii, 68, 100, 102, 126, 134, 137

Hourani, George, 141n3

How Anyone Can Become a Philosopher on His Own (Abubachar), 71, 153n15

Huntington, Robert, 108, 113, 165n34, 169n62

Huygens, Christian, 10, 137

Ibn al-'Arīf, Aḥmad ibn Muḥammad, 25–26, 142n20

Ibn 'Ali, Tāshufīn (sultan), 26

Ibn al-Nafī, 'Alī ibn Abī al-Ḥazm, 41

Ibn-Bājja, Abu Bakr Muhammad ibn Yahya. See Avempace

Ibn Barrajān, 'Abn al-Salām, 25–26, 142n20

Ibn Bashkuwāl, Khalaf ibn 'Abd al-Malik, 144n39

Ibn-'Ezra, Avraham, 156n31

Ibn-Ḥamdīn, Husayn, 16, 19, 21, 22, 30, 144n35

Ibn Qasi, Abū al-Qāsim, 25, 26, 142n20

Ibn al-Qaṭṭān al-Fāsī, 'Alī ibn Muḥammad, 26

Ibn al-Qunfudh, Aḥmad ibn Ḥusayn, 21
Ibn-Rushd, Abū 'l-Walīd Muḥammad. *See* Averroes
Ibn Rushd al-Jadd (grandfather of Averroes), 29, 144n35
Ibn-Tibbon, Shmuel, 48–49, 57–58
Ibn-Tibbon, Yehuda, 149n15
Ibn-Tufayl, Abū Bakr Muhammad (Abubachar Eben Tofail), 2; and Almohads, 26–30; on autodidacticism, vii, 9, 16, 21–22, 38–39, 40, 59; and Avempace, 30; background of, 16; and al-Ghazzālī, 15, 21–23; on God, 22, 24; influence of, 10; on knowledge, 23–25; al-Marrākushī on, 27–28; and Narbonni, 57, 153n13; and Pico, 71, 75; Pococke on, 105; and religion, 119; sources of, 129; and spontaneous generation, 116, 118; *vs.* Sufism, 13, 23, 24
Ibn-Tūmart, Muhammad, 16, 21, 25–30, 28, 40, 143nn26,30
Ibn Wakar, Joseph, 148n7
The Idea of Robinson Crusoe (Pastor), 10
Idel, Moshe, 150n44, 155n27
al-Idrīsī, Abū 'Abdallāh, 34, 36
al-Ilbīrī, Muhammad b. Khalf b. Mūsā al-Anṣārī, 144n35
Immediate Revelation (Keith), 168n58
The Improvement of Human Reason Exhibited in the Life of Hai Ebn Yokdan (trans. *Ḥayy Ibn-Yaqẓān*; Ockley), 10, 12
Indian Ocean, 32–34, 36, 40, 133
The Infancy of Pico (painting; Delaroche), 68, 69
Innocent VIII, Pope, 65, 67
Instauration magna (*The Great Instauration*; Bacon), vii, 99
islands, desert, 31–35, 40, 124–125, 132–133, 135, 137, 140n13. *See also* Wāqwāq, island of

James I (king of Aragon), 43
James I (king of England), 106
Jews, 29, 73, 76, 110, 146n4, 149n15, 164n24; in Barcelona, 61–62; and Christians, 42–43; and education, 43–44, 129; excommunications by, 52–53, 56, 57, 59, 61, 62, 63, 129–30; and *Ḥayy*, 41, 129; laws of, 53, 54; persecution of, 46, 60; *vs.* philosophy, 42, 46–56, 62; and trade, 35, 43. *See also* God
The Jungle Book (Kipling), 2. *See also* Mowgli

Kabbalah, 49, 74, 85, 150n44, 152n11, 170n6; in *Ḥayy*, 88–89; and Pico, 67, 73, 75, 76, 77, 86, 154nn17,18, 156n34
Kalila wa dimna (*Fables of Bidpai*), 139n7
Kant, Immanuel, 126
Kashlari, Avraham, 148n7
Keith, George, 10, 119–20, 168n58
al-Khwārizmī, Muhammad ibn Mūsā, 34
Kitāb al-Ḥadaiq (al-Baṭalyawsī), 155n27
knowledge, 18, 27, 115, 127, 142n9, 154n21; and culture, 9, 102; desire based on, 87; and experience, 19, 21–22; of God, 21, 22, 24, 28, 105, 111, 119, 125, 141n3; Grotius on, 113–14; in *Ḥayy*, 23–25, 89, 129; hierarchy of, 23–25, 58, 105; and mysticism, 31, 102, of nature, 2–3, 21–23, 90, 125; and religion, 119–20; Sufis on, 21–22; theories of, 28, 31, 70, 102, 141n3
Kriegel, Maurice, 146n4

Ladder of Spiritual Life (illustration; Rosselli), 85
Landucci, Luca, 96
Laplanche, François, 111
Latin language, 42, 64, 129
Laud, William, 107, 109, 110, 163n12
Leibniz, Gottfried Wilhelm, Freiherr von, 10, 137
Lemnius, Levinus, 167n50
Leonardo da Vinci, 82
Leoniceno, Niccolò, 87, 154n16
Lesley, Arthur, 155n27
A Letter on the Perfection of the Soul (Narbonni), 148n6
Lévi-Strauss, Claude, 39
Libre de homine (Manfredi), 86
Life of Pico (*Vita de Pico*; Gianfrancesco Pico della Mirandola), 93, 99, 131
Lincei Society, 99, 134, 135
Locke, John, 108, 119, 137, 169n62; on autodidacticism, vii, 5–6, 121, 123–24; on blank slate, 9, 120–24; and Pococke, 10, 12, 120, 121, 123–24, 161n6, 169nn64,68
Lucian, 131
Lucretius, 118, 167n47

al-Ma'ārif al-'aqlīyah (al-Ghazzālī), 15, 22
Machiavelli, Niccolò, 68, 80
Madonna of the Steps (painting; Michelangelo), 85
Magia naturalis (della Porta), 131

magic, 53, 55; natural, 76, 87, 99, 130, 131, 152n11, 156nn34,35; and Pico, 67, 76–77, 87, 156n34
Maimonides, Moses, 44–46, 58, 62, 111, 146n1, 147n5, 150n42; and Jacob's ladder, 48; and Locke, 120–21; Narbonni's commentary on, 56, 63, 148nn6,11; and pedagogical controversy, 53, 54
Mainardi, Giovanni, 78
Mālikī school (Islamic law), 26
Manfredi, Girolamo, 86, 158n65
Manilius, Marcus, 159n74
Mari of Lunel, Aba, 50–52, 53, 54, 62
al-Marrākushī, 'Abd al-Wāhid, 25, 27–28, 29, 30
al-Mas'ūdi, Abu al-Ḥasan 'Alī, 31–32
mathematics, 3, 22, 26, 28, 31, 40, 48, 53, 112, 120
Meadows of Gold (*Kitāb murūj al-dhahab*; al-Mas'ūdi), 31
Medici, Lorenzo de': death of, 66, 70, 78–80, 88, 93, 130; and homosexuality, 82, 83; philosophy of, 87, 89; and Pico, 65, 68, 77, 156n37; and Savonarola, 95
Medici, Piero de', 157n59
Meteorology (Aristotle), 34
Michelangelo, 85
Michelozzi, Niccolò, 93
Mill, John Stuart, 138
Miller, Larry, 147n5
Miraculous Conformist (Stubbe), 113
Mitchell, James, 139n3
Mithridates, Flavius (Shmuel Ben-Nissim Abu al-Faraj), 154n18, 155n28
Moody, Ernest, 116
More, Henry, 168n58
More, Thomas, 99, 102, 127, 131–33, 161n1
The Most Precious Thing Asked For (*A'azz mā yuṭlab*; Ibn-Tūmart), 26
Mowgli (*Jungle Book*), 2, 4, 138
Murād III (Ottoman sultan), 105–6
mysticism, 40, 128, 131, 135; Eastern, 31, 33–34; illuminist, 31, 119–20, 142n3; and knowledge, 31, 102

Nahmanides, 150n42
Narbonni, Moses, 42–64, 129, 130; and Alemanno, 73, 74; and autodidacticism, 42, 57, 59–60, 62, 151n46; and Averroes, 44, 60, 147nn5,6; on God, 44, 46, 57, 58, 60, 61, 62, 147n5, 151n46; life of, 44–46; on Maimonides, 56, 63, 148nn6,11; and pedagogical controversy, 54–56, 57–59; and Pico, 65, 86; and Pococke, 110. See also *Yehiel Ben-'Uriel*
nature: active *vs.* contemplative approach to, 90, 129; alive *vs.* dead, 89–93, 127, 128; and autodidacticism, 59, 126; and Avempace, 30; Bacon on, 99–100, 135; and children, 4–9; *vs.* civilization, 2, 48, 112; conquest of, 36, 65, 99, 127; control of, 127–28, 152n11; coupling with, 70, 75; exploration of, 28, 40–41, 46, 58, 61, 63, 75, 92, 104–5, 130, 132, 137; and *Ḥayy*, 23, 28, 89, 128; knowledge of, 2–3, 21–23, 90, 125; in ladder of teaching, 58; and natural magic, 76; in painting, 65; Pico on, 65, 75, 76, 86, 87, 89–93; and Pococke, 101, 108–9; and reason, 2, 27; and religion, 52, 118–20; and totemic taxonomy, 39
The nature of the drink kauhi, or coffe (Antaki), 111
Der Naturmench (The natural man; trans. *Ḥayy Ibn-Yaqẓān*; Eichhorn), 10
Neoplatonism, 82, 84–86
New Atlantis (Bacon), 127, 135–36, 170n6
Newbery, John, 6
New Experiments Physico-Mechanicall, Touching the Spring of the Air, and its Effects (Boyle), 116, 117
News from Aleppo (Robson), 106
Nicomachean Ethics (Aristotle), 89
Nogarola, Leonardo, 66, 152n4

Ockley, Simon, 10, 12, 162n6
Oldenburg, Henry, 168n53
Opere (Fregoso), 97
Oration On the Dignity of Man (Pico della Mirandola), 67, 75, 76, 86, 108, 152n11
Ordelaffi, Pino, 158n65
L'Orfeo (play; Poliziano), 83
orientalism, 105, 109, 161n6
Ottoman Empire, 105–6
Oxford University, 13, 101, 102, 105, 106, 110, 124

Pagnini, Sante, 155n28
Panchatantra, 4
Partlicius, Simeon, 166n41
Pasor, Matthias, 106
Pastor, Antonio, 10, 140n13
Paul V, Pope, 131
pedagogical controversy, 42, 43–44, 46, 51–56, 59, 61, 62, 129–130

Pedro IV (king of Aragon), 56
Philo, 102
Philosophical Transactions of the Royal Society (journal), 103
Philosophus autodidactus (trans. Ḥayy Ibn-Yaqẓān; Pococke), 15, 101–5, 119, 137; and Boyle, 116, 118; historiography of, 161n6; and Locke, 10, 12, 120, 121, 123–24, 161n6; title of, 102
philosophy: and active *vs.* contemplative life, 48–49, 57, 58, 90, 127, 129; adolescents' study of, 42, 43–44, 46, 51–56, 59, 61, 62, 129–30; Andalusian, 15, 22, 28, 30, 31, 40, 116, 128–29, 138; Arabic, 48, 49; *vs.* authority, 51–53; Greek, 49, 55; illuminist, 31, 119–20, 142n3; *vs.* Jewish law, 42, 46–56, 62; ladder of, 58, 153n11; and mysticism, 40; and Pico, 87; popularization of, 50–52, 63; women's study of, 52, 53. *See also* Aristotle; Plato; Socrates
Phosphorus, Lucio, 68
Physics (Aristotle), 144n40
Pico della Mirandola, Gianfrancesco, 78, 93, 99, 131
Pico della Mirandola, Giovanni, 10, 65–100, 135, 137; on Adam, 75, 130; on astrology, 65, 66, 71, 76–78, 88, 90, 91, 92, 95, 97, 152n11, 155n30; *vs.* authority, 66, 68, 71, 153n11; and autodidacticism, 68, 69, 75, 77, 92; and Bible, 73, 86; and Boyle, 117; on climate, 91–92; on experience, 65, 71, 95; and homosexuality, 81, 84, 86–89, 158n61; influence of, 99–100; and Kabbalah, 67, 73, 75, 76, 77, 86, 154nn17,18, 156n34; life of, 66–67; *Life* of, 93, 99, 131; and Lorenzo de' Medici, 65, 68, 77, 156n37; loves of, 86–89; and magic, 67, 76–77, 87, 156n34; murder of, 93, 151n2; *Oration* of, 67, 75, 76, 86, 108, 152n11; poetry of, 88; portrait of, 94; *vs.* predestination, 71, 73, 76, 90; and Savonarola, 78, 93–95, 96–97; on spontaneous generation, 73, 91–92; *Theses* of, 67, 76, 77; translation of Ḥayy by, 71–76, 88–89, 97. *See also Disputationes adversus astrologiam divinatricem*
Pindar, Paul, 163n10
Pinker, Steven, 139n6
Plato, 24, 30, 34, 38, 87, 89, 126
Platonic love, 66, 70, 86–89, 93, 130
Platonic Theology (*Theologia Platonica de immortalitate animorum*; Ficino), 90
Pliny, 108
Pococke, Edward, 10, 20, 101–5, 141n3; and autodidacticism, 15, 101, 102, 105, 124; and Boyle, 111, 112–16, 163n14, 164n27; career of, 105–11; and chameleon, 101, 108–9; and experimentalism, 101, 102–5, 111, 124, 137, 161n6; and Locke, 120–21, 123; manuscript collecting by, 107–8; and solitude, 101, 102, 108, 111, 124, 137; translation of Grotius by, 111–14. *See also Philosophus autodidactus*
Pococke, Edward, Jr., 169nn64,68
poetry, 88, 97–99, 109, 131
Poliziano, Angelo, 65, 68, 78–80, 87, 88, 89; and homosexuality, 82, 83, 84; murder of, 93, 151n2, 160n90
Porta Mosis (Pococke), 111, 120–21
Possibility of Conjunction (Averroes), 60
The Possibility of Union (Efsharut haDevekut; Averroes), 44
The Praise of the Folly (Erasmus of Rotterdam), 131
predestination, 71, 73, 76, 90, 95
Principles of Cartesian Philosophy (Spinoza), 10
Pritsius, Georg, 10
Ptolemy, 34–35, 134
Pyro, 118, 167n47
Pythagoras, 34, 77, 87, 134

Quakers, 119–20, 168n58
Quran, 26, 27

The Red Violin (film), 14
Refutation (Ibn-Ḥamdīn), 19, 21
The Regime of Solitude (Avempace), 46, 58, 144n40, 148nn6,11, 150n38
Regiomontanus, Johannes, 159n74
Renan, Ernest, 41
Republic (Plato), 24
The Revival of the Religious Sciences (*Ilḥyā' 'ulūm al-dīn*; al-Ghazzālī), 16–19, 21–23, 129, 144n35; suppression of, 25–26, 29
Robinson Crusoe (Defoe), 1, 10, 140n13, 161n6
Robson, Charles, 106
Rocke, Michael, 81
Romano, Jacob, 110
Roselli del Turco, 76
Rosselli, Francesco, 85
Rousseau, Jean-Jacques, 6, 7, 9, 12, 137–38
Russell, G. A., 161n6
Rutkin, Darrell, 76, 155n29

Sacrifice of Zeal (*Minhat kanaut*), 50
Saltarelli, Jacopo, 82
Saltness of the Sea (Boyle), 165n34
Sandford and Merton (Day), 6, 9
Saturnia regna (The kingdom of Saturn; Virgil), 127
The Savage Mind (*La pensée sauvage*; Lévi-Strauss), 39
Savonarola, Girolamo, 66, 70, 79, 130, 151n3; on climate and politics, 95–96; eulogy for Pico by, 93–94; execution of, 93; and Pico, 78, 93–97; and Pico's murder, 95; portrait of, 94
Scala della vita spirituale sopra il nome di Maria (Domenico Benivieni), 85
Seaman, William, 165n27
Sefer Abu-Baker: Hayy Ben-Yoktan (Narbonni), 47
Sennert, Daniel, 167n50
sexuality, 78, 88, 89. *See also* homosexuality
Sforza, Lodovico, 96
Simoneta, Barolomeo, 160n100
Simoneta, Bonifacio, 160n100
Simoneta, Giovanni, 160n100
Sirigatti, Antonio, 156n31
Socrates, 84, 93
solitude, 1, 59, 115; and autodidacticism, 30, 49, 62; in *Ḥayy*, 88, 129; and Narbonni, 46, 147n5, 151n46; and Pococke, 101, 102, 108, 111, 124, 137
Somenzi, Paolo de, 96
Some Thoughts Concerning Education (Locke), 5, 123–24, 137
Song of Songs (Bible), 48–49, 57, 73, 86
Spinoza, Baruch, 10, 63, 137
da Spoleto, Pierleone, 76, 78, 80, 157n47, 159n74
spontaneous generation, 31–32, 36, 40, 129; and experimentalism, 116, 124, 125; Fregoso on, 97–98; mechanical, 116–18; Pico on, 73, 91–92
Starkey, George, 166n41
Stoicism, 115
Strozzi, Tito Vespasiano, 67
Stubbe, Henry, 113
sudden perception, 59–61, 130, 150n44, 151n46
Sufism, 129; and Almohads, 25–31; and authority, 15, 16, 19, 25, 30, 144n39; and al-Ghazālī, 17, 19–23, 25–26, 28; God in, 15, 19, 21, 22, 40; vs. Ibn-Tufayl, 13, 23, 24; and revolts, 142n20; self-discovery in, 15–17; and trade networks, 33–34

Suharawardī, Shihāb al-Dīn, 141n3
Sylua syularum, 136

Taprobane (Sri Lanka), 35, 133
Tarzan (Burroughs), 2, 4, 138
Theologus Autodidactus (Ibn al-Nafīs), 41
Timaeus (Plato), 34
Tom Telescope series, 6
Toomer, Gerald J., 161n6
totems, 38–40
trade: British, 105–6, 107, 124, 162n10, 165n34; cross-cultural, 162n10; with Far East, 32–34, 35; and Jews, 35, 43; and lost islands, 31–35; and Pococke, 105, 124; and Sufis, 33–34
Trimmer, Sarah, 9
Twells, Leonard, 106

Unitarianism, 118
Ussher, James, 108
Utopia (More), 99, 102, 127, 131–33, 132
utopianism, 10, 15, 99, 101, 111; and autodidacticism, 126–28; and Boyle, 115–16; and desert islands, 31–35, 40, 124–25, 132–33, 135, 137, 140n13; in England, 131–33; and experimentalism, 124; Greek, 126–27, 128

Valeriano, Pierio, 93
Vidal, Bonfosh, 53
Vidal, Don Crescas, 54–55
Vidal, Pierre, 148n8
Virgil, 127
Vishnu Sarma (*Panchatantra*), 4
Voltaire, 10
Der von sich selbst gelehrte Weltweise (The self-taught world wise; trans. *Ḥayy Ibn-Yaqẓān*; Pritsius), 10

Wallis, John, 110
Walton, Brian, 108
Wāqwāq, island of, 31–35, 40, 57, 58, 91, 118, 129, 133; women on, 35–37, 145n56
Wilkins, John, 110
Wirszubski, Chaim, 154n18
women, 9, 35–37, 52, 53, 124, 145n56, 152n4
Work for Chimny-Sweepers or A Warning for Tabacconists (Philaretus), 166n41
Wyche, Peter, 110

Yates, Frances, 152n11
Yehiel Ben-ʿUriel (trans. Ḥayy Ibn-Yaqẓān; Narbonni), 42, 46, 55–64, 129, 130, 150n38; illustrations from, 55; and pedagogical controversy, 57–59; and Pico, 65, 73, 74; title of, 57, 62; translations of, 64, 65, 146n2

Zoroaster, 77